Bluntly Said

CULTURE RE-WIRED

UNLEASH YOUR INNER AI CEO

IDA BYRD-HILL

Bluntly Said
An Imprint of Automation Workz, Inc.
P.O. Box 930709
Wixom, MI 48393-0709

Published by Automation Workz, Inc.

For Information about special discounts for bulk purchases and speaking engagements

Please contact 313-444-4885 or business@autoworkz.org

Name:	Ida Byrd-Hill		
Title:	Culture Re-Wired		
	Unleash Your Inner AI CEO		

Subjects:	Business	Artificial Intelligence	Human Resources

ISBN 979-8-9997461-0-8

Table of Contents

Introduction

I've been called "blunt" my whole life—sometimes as a compliment, more often as a critique. I call it being efficient. Why waste laps circling the track when you can take the inside line straight to the point? That no-nonsense style is exactly why I'm launching my new book series, **Bluntly Said**—a high-octane mash-up of business culture, behavioral economics, human resources, workforce development, and technology transformation. No fluff. No detours. Just the real roadmap to change. *Culture Re-Wired: Unleash Your Inner AI CEO* is the first book off the line in this hard-hitting series—and trust me, it's built for speed.

I've watched front-liners up close—white-knuckling the wheel of life while navigating a grueling full-time physical job, high-octane tech coursework, and the constant traffic jam of family responsibilities—all with one destination in mind: breaking free from the poverty lane. Yet too often, the world assumes they're idling in neutral content to coast in the breakdown lane of low wages. *The truth?*

These drivers of the everyday economy are hungry for the fast lane, aiming squarely for tech careers. They see the warning lights flashing—automation is overtaking the manual jobs they hold today—and they're ready to shift gears. But far too often, they feel like invisible passengers, never handed the keys to opportunities that would boost their skills, income, and status.

Culture Re-Wired: Unleash Your Inner AI CEO is my pit crew manual for closing that gap between executives, middle managers, and front-liners—because in today's AI Grand Prix, every city, state, and nation needs a full team on the track. Real transformation requires every driver, mechanic, and strategist in the race.

To turbocharge these pages, I packed in corporate case studies that show the principles in action. And, true to my love for technology, I didn't just fuel up on traditional research—I hit the nitrous. With AI chatbots like ChatGPT, Claude, Gemini, Grok, and Perplexity in my digital garage, I scoured annual reports, news archives, and journal articles at record speed. What once would have been a 1,100-hour marathon became a 300-hour sprint, shaving off 800 hours and proving

AI isn't just a passenger—it's the engine. **My hope?** That these case studies ignite your team's engines and push you to unleash your inner AI CEO to win this global AI race.

Culture Re-Wired: Unleash Your Inner AI CEO

Chapter 1: Welcome to the Wild, Wild AI Age

Newsflash: artificial intelligence isn't tiptoeing toward the mainstream — it just hijacked the calendar and dragged tomorrow into today. Algorithms now steer forklifts through warehouses, crank out digital ads at hyperspeed, and make your phone feel downright telepathic. Meanwhile, *generative* AI— the sorcerer that can spit jokes, sketch

Illustration Automation Workz

logos, and write production-ready code in a heartbeat—has executives acting like they've stumbled onto a gold rush on Mars.

From Crystal-Ball Dreams to Hard-Cash Reality

The impact of AI on profits by industry

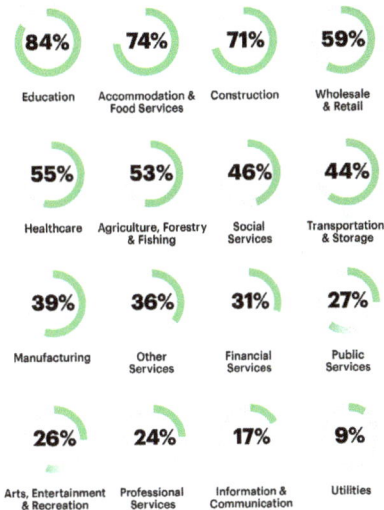

Industry	%
Education	84%
Accommodation & Food Services	74%
Construction	71%
Wholesale & Retail	59%
Healthcare	55%
Agriculture, Forestry & Fishing	53%
Social Services	46%
Transportation & Storage	44%
Manufacturing	39%
Other Services	36%
Financial Services	31%
Public Services	27%
Arts, Entertainment & Recreation	26%
Professional Services	24%
Information & Communication	17%
Utilities	9%

2017, Accenture

Back in 2017, Accenture raised eyebrows with a bold forecast: AI would spike profits by an **average 38 percent** across sixteen industries by 2035. The early 2024 numbers reveal a crowded scoreboard of sprinters and stragglers:

Manufacturing is already lapping predictions. With robot vision, generative design, and predictive maintenance, plant-floor AI rocketed to a **$3.5 billion** niche in 2023. It is expected to be a **$58 billion** giant by 2030 — nearly **48%** CAGR (Compound Annual Growth Rate).

Financial services?
Printing money. Bank of America estimates AI could add two full points to profit margins within five years — small on paper, massive on a trillion-dollar balance sheet, amounting

to $139.5 billion.

Healthcare and **education** possess sky-high potential but remain tethered to red tape and privacy landmines; their takeoff has been more paper plane than jet engine, so far.

Half of early adopters already expect to **double** their AI investment in just three years, while front-line productivity jumps by **20–30 percent**. Wages in AI-heavy jobs are soaring twice as fast as in low-automation roles—proof the talent wars have begun.

The investment cash floodgates? Wide open. The 2025 **Stanford AI Index** reads like a financial thriller on turbo-boost:

Statistics	Why It's Jaw-Dropping
252 Billion poured into corporate AI in 2024—**26% up** in a single year, **13** times 2014's tally	That's GDP-level spending on code and silicon.
$34 Billion funneled into *generative* AI alone—nearly **9** times 2022, now **20 %** of all AI dollars	Everyone wants the magic idea machine.
$109 Billion U.S. private AI haul—**12** times China's spend	Uncle Sam just hit the afterburners.
78% of companies using *some* AI—up from 55 % last year	Resistance is officially futile.
71% have plugged gen-AI into at least one workflow—was 33 % a year ago	Adoption curve? More like a rocket-launch graph.
27 K industrial robots China installed in 2023—more than the rest of the planet combined	Assembly lines are turning into sci-fi sets.
10 % of new bots are "cobots," built to work shoulder-to-shoulder with humans	Your next cubicle buddy might have gears.

AI is no longer a showroom toy. It's the new workhorse, already hauling in savings, birthing fresh revenue streams, and lapping competitors still fiddling with spreadsheets. The race has started; first movers are pulling ahead. Hitch your wagon to this horsepower now or get comfortable eating dust.

Generative AI: The Profit Turbocharger
Forget incremental gains—this is the digital super-charger that slams the profit pedal through the floorboards. Why are seasoned CEOs whispering about gen-AI like it's a secret mint? Because it:

- Blitzes out emails, blogs, and ads in seconds—copywriter dream team at 1,000 × speed.

- Handles customer chats 24/7 with zero eye-rolling, no mood swings or hold music.

- Reads data, images, even emojis to predict what buyers want *next*.

- Vaporizes paperwork so employees can tackle projects that actually fatten the P&L (Profit & Loss) statement.

- Brainstorms new products nonstop, fueling a constant launch cadence.

- Never calls in sick — revenue keeps humming while humans sleep.

Strap this turbocharger onto your business engine, and rivals will only see tail-lights.

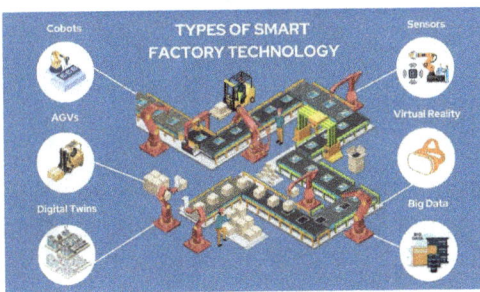

Illustration Automation Workz

Industry 4.0: Generative AI at Full Throttle
Generative AI isn't a bolt-on gadget; it's rocket fuel, super-glue, and GPS for the entire Industry 4.0 juggernaut. Picture fleets of smart manufacturing factories where robots learn in real time, IoT

devices gossip nonstop, digital twins solve glitches before the night shift clocks in. Generative AI is the ringleader electrifying every bolt and circuit.

Profits That Pop Like Champagne

Early adopters are already bragging about **10%–38%** jumps in profit margins after unleashing generative AI on everything from design to quality control. Analysts forecast this single technology will mint $4.4 TRILLION in yearly value by 2030—and that tally *doesn't* include the wave of Industry 4.0 spinoffs (edge analytics, autonomous cobots, predictive supply chains) it's about to turbocharge.

Hospitals: Smart Factories of Human Longevity

Healthcare isn't just catching up; it's rewiring. Think of hospitals as **"factories of life"** pressing out extra miles of vitality for 70 million baby boomers looking to swap worn-out parts for extended-warranty bodies. AI—the new master engineer keeping every conveyor belt of care humming at warp speed.

Step inside the hospital of yesterday and you'd find a bureaucratic maze: an MRI appointment slotted for "next week," dog-eared paper charts stuffed into metal folders, and a reactive care model that sprang into action only after the patient's condition sounded alarm bells. Rehabilitation was a one-size-fits-all regimen—walk the hallway, count your steps, hope for the best.

Now fast-forward to **Smart Hospital 4.0**, a high-octane command center where AI directs every heartbeat of care. The moment you check in, an AI triage system reviews your symptoms, cross-checks them against millions of prior cases, and bumps you straight to imaging *today*—all while drafting the radiology report before you reach the scanner. Paper charts are relics; instead,

Illustration Automation Workz

digital twins of your organs run virtual drug trials in seconds, revealing how each therapy will play out before a single pill ever touches your tongue.

Care is no longer reactive. Predictive algorithms comb through real-time vitals, lifestyle data, and genetic markers, flagging a brewing cardiac episode months before it can ambush you. Rehabilitation, too, has gone bespoke: AI-powered exoskeletons learn your gait in real time, adjusting support with every shaky step until you stride out stronger than before.

Hovering over it all are robotic arms armed with what amounts to centuries of aggregated surgical expertise, chatbots that monitor your vitals at 2 a.m. like tireless night-shift nurses, and pharmacy algorithms concocting custom drug cocktails in hours instead of months. The result? Costs plunge, outcomes soar, and life expectancy transforms from a hopeful aspiration into a scalable product line—factory-precision longevity for the modern age.

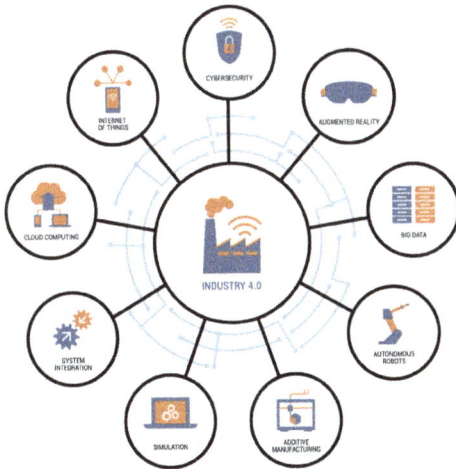

Illustration Automation Workz

Let's look at some of the technology, led by AI, that is impacting _all_ smart factories.

Autonomous Robots
Meet the metal masterminds rolling off assembly lines and zipping through warehouses. With AI as their turbo-charged cortex, these bots don't just follow orders—they read the room. Cameras catch a raised eyebrow, sensors notice a shaky hand, and in a millisecond the robot shifts course: gentler grip, wider berth, zero drama.

Additive Manufacturing (3-D Printing)
Sketch it at lunch, hold it by dinner. AI-driven printers slice wild ideas into feather-light lattices and bullet-proof frameworks, spraying molten magic by the layer. Yesterday's "impossible" parts become today's show-stopping prototypes.

Simulations & Video Games
Strap on the headset and step into the proving ground. AI-fueled worlds push recruits, pilots, and managers through crisis after crisis — every click, dodge, and decision logged. Hidden leaders surface, sleepy skills awaken, and talent pipelines fill themselves.

System Integration

Picture every gadget in your tech zoo finally speaking the same language. AI writes the on-the-fly Rosetta Stone, welding apps, sensors, and machines into one swaggering super-system that never drops the ball.

Cloud Computing

Forget server closets that howl like jet engines. AI now flexes limitless cloud muscle—crunching petabytes in minutes, scaling costs down to pocket change, and leaving old-school hardware in the dust.

Internet of Things (IoT)

From thermostats to 18-wheelers, billions of gizmos gossip nonstop. AI sifts through the chatter, predicts a hiccup before it happens, and tunes operations so tight they practically hum.

Cybersecurity

Digital ninjas never sleep. AI hunts anomalies at warp speed, slams the gates on phishing phantoms, and foils ransomware heists before ransom notes hit the inbox.

Augmented Reality (AR)

Wave your phone and watch reality get an upgrade. AI paints holographic instructions over machinery, senses your frown of confusion, and cues an instant walkthrough—like having a genius mentor hovering in thin air.

AI is the rocket booster of Industry 4.0 — scanning, thinking, creating and seizing opportunities 24/7 to accelerate innovation, speed workflows, slash costs, and generate fresh revenue streams. Analysts peg the payoff at a staggering $22 trillion a year by 2030—money on the table for anyone bold enough to grab it.

Translation: no sector, no leader, no corner office (or shop floor) gets a free pass. If you are employed, you are

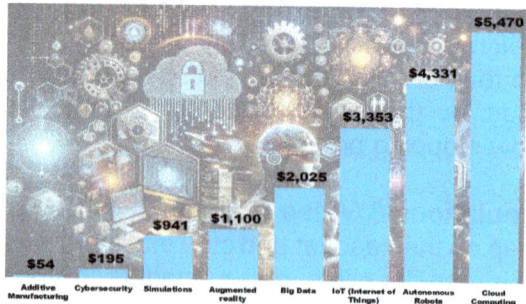

EMERGING TECHNOLOGY PROJECTED REVENUE BY 2030
In Millions USD

Additive Manufacturing	$54
Cybersecurity	$195
Simulations	$941
Augmented reality	$1,100
Big Data	$2,025
IoT (Internet of Things)	$3,353
Autonomous Robots	$4,331
Cloud Computing	$5,470

PROJECTIONS 2024-2030 SOURCE: Open AI (Chat GPT) PHOTO:Open AI (ChatGPT riding shotgun on the digital transformation superhighway — so buckle up and press the pedal to the metal and take off.

Pandemic Profits: The Shockwave That Turbo-Charged AI

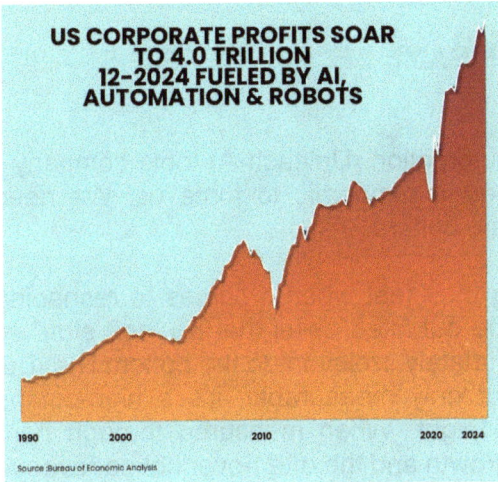

US CORPORATE PROFITS SOAR TO 4.0 TRILLION 12–2024 FUELED BY AI, AUTOMATION & ROBOTS

Source: Bureau of Economic Analysis

Picture this: 2021—streets half-empty, supply chains snarled and Zoom fatigue at work. Yet behind boardroom doors, profit meters were spinning like slot machines on a hot streak. Corporate America roared to a jaw-dropping $2.8 trillion in after-tax earnings — the fattest corporate profits since Harry Truman was in the White House! How did companies cash in while the world pressed pause?

Three seismic forces collided to create a once-in-a-century money-maker:

Front-Liners Became the Unbreakable Backbone
Grocery clerks, delivery drivers, factory operators — everyday heroes who refused to let the economy flatline. Their grit kept goods flowing and shelves stocked, turning "essential" into "irreplaceable."

Robots Grabbed the Steering Wheel
Labor shortages? No problem. Businesses unleashed armies of warehouse bots, self-checkout lanes, and contactless delivery drones. Jobs that once took five humans were hammered out by one worker and a fleet of silicon sidekicks.

AI Transformed Hustle into Hyper-Efficiency
Chatbots soothed anxious customers 24/7. Predictive engines forecast inventory needs before managers finished their coffee. Algorithms slashed waste, squeezed margins—and never called in sick.

The result? A productivity boom, so ferocious, it rocketed corporate profits to a record $4 trillion by 2024. Humans and machines formed an unstoppable tag team, proving that in the face of crisis, innovation isn't optional—it's rocket fuel. McKinsey & Company also noted that businesses embracing digital automation during the pandemic outperformed their competitors in both efficiency and earnings.

Now CEOs are asking: **"How do we recreate that same profit — without the chaos of a crisis?"**

The answer is clear: USE THE SAME SECRET WEAPONS — BUT DO IT PROACTIVELY.

Empower front-liners. Scale automation. Unleash AI tools company - wide. This time, we don't need a pandemic to force us. We need leadership bold enough to rewire culture.

Far too many executives drag their feet when it comes to reshaping company culture, clinging to the outdated belief that it's "soft stuff" — elusive, immeasurable, and ultimately irrelevant to the bottom line. But here's the truth: culture is not only measurable, it's a high-octane growth engine hiding in plain sight. When measured through hard financial metrics like revenue growth and the often-overlooked revenue per employee, culture reveals itself as a decisive competitive weapon. Companies that get culture right don't just see incremental improvements — they unleash performance surges that generate new revenue while energizing front-liners to retain current revenue.

Meet the New Heroes of the AI Age
Ditch the clichés — today's trailblazers don't sport capes; they rock aprons, scrubs, and steel-toe boots. The real architects of the AI uprising aren't cloistered in Silicon Valley — they're pouring lattes, stocking shelves, tuning engines, caring for patients, and wrangling pallets. These everyday heroes wield a new superpower: harnessing tech to think bolder, work sharper, and ignite change from the shop floor up.

In these pages, you'll tour everything from mom-and-pop coffeehouses to family-run plants and Fortune 500 titans — and you'll spot the common spark: **culture comes first.** Before they downloaded a single algorithm, these companies re-wired their cultures to turbo-charge their people — C-suite strategists, middle-lane managers, and front-line dynamos alike. They didn't just "install" AI; they unleashed it by empowering the very hands that hit the keys and turn the wrenches. Now front-liners armed with generative AI are re-engineering daily routines, spiking productivity, and driving profits sky-high — one barcode scan, one patient chart, one espresso shot at a time.

This book unveils the **Inner AI CEO Mindset** — a playbook that turns every apron-wearing cashier, floor-walking supervisor, and corner-office power broker into a profit-commanding trailblazer. Real CEOs don't wait for orders; they set the agenda, seize initiative, and magnetize the crowd. Infuse that stance with emotional intelligence, off-the-charts creativity, crystal-clear communication, and laser-cut critical thinking, and AI morphs from shiny tool to unstoppable force multiplier.

You are the AI Culture Architect. Algorithms can crunch data at warp speed, but they can't build trust, spark passion, or shift mindsets. Only humans can. That's your cue. You don't need a gold-plated title or a Ph.D. in code—just the guts to ask:

"What if we tried this?"
"What's a better way?"
"How can we use AI to make workplace culture more human?"

Buckle up — those questions just flipped the ignition on **Culture Re-Wired**, launching you and your team straight into **Inner AI CEO** mode. And here's the turbo-charged bonus: this culture overhaul doesn't just fatten the company's bottom line; it can catapult *your* career trajectory and paycheck — if you're clocking every cost slash and profit spike along the way.

Think of this book as a high-velocity GPS for maximum profitability. Like smart traffic lights that keep the freeway humming, there are signs — mindset, metrics, tips — so your transformation accelerates without a single bottleneck.

Pay particular attention to the icons especially the green graph.

❌   Red icons are items that need to be stopped.

✅   Green are icons that need to be implemented.

This icon is a profit-making action to improve the corporate bottom line.

Let's begin!

Chapter 2: Culture Drives AI, Not the Other Way Around

Corporate culture isn't some feel-good poster on the break-room wall — it's the invisible engine revving (or stalling) every dollar your company makes.

Culture is the office atmosphere — literally. It's the invisible oxygen everyone inhales between answering emails and microwaved leftovers. Culture is the hush-hush code of conduct people follow when the boss is off "working remotely" (a.k.a. golfing). It decides whether folks raise their hands with wacky ideas or keep them glued to the keyboard in silent terror; whether bug reports are treated like science experiments or public executions; and whether mistakes become catapults for growth or catapults for — well — people.

Spoiler alert: culture is **not** the free pizza you bribe the interns with or that inspirational poster of a lone eagle soaring over a canyon of KPIs. It's the unseen bedrock that separates AI gold mines from AI money pits. Culture shows up in the tiniest moments — like whether someone feels safe enough to ask a "dumb" question, admit a blunder, or experiment with a shiny new tool.

Generative AI needs a workplace where employees are mentally, emotionally, and socially primed to adopt and scale new tech. That's the difference between tiptoeing into the future and sprinting toward it with jet packs.

Culture: The Force That Makes — or Breaks — Millions
Picture a workplace so electric that every morning feels like the launch-pad countdown at Cape Canaveral. The moment teams log in, engagement rockets sky-high — targets aren't merely met; they're obliterated, then celebrated, then shattered all over again as employees stick around for the next encore performance. In this adrenaline-charged atmosphere, collaboration isn't a buzzword; it's jet fuel. Ideas ricochet across departments, collide, and fuse into innovations that send digital-transformation timelines into warp speed.

The outside world can't help but notice. Your employer brand becomes so magnetic it practically bends LinkedIn's algorithm: A-list talent floods your inbox, signs offer letters in record time, and — miracle of miracles — actually shows up on day one, pumped to contribute. And when employees feel this fired-up, customers ride the same high.

Service soars to five-star heights, loyalty locks in like titanium, and repeat sales glide in on autopilot. In short, when engagement, innovation, branding, and customer delight hit this stratosphere, your company isn't just playing the game — it's rewriting the scoreboard to maximum profits.

Why Front-Liners Hold the Culture Keys

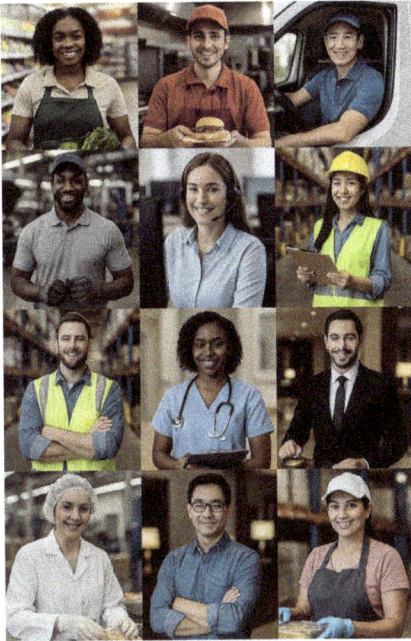

Front-liners — cashiers, agents, technicians, nurses, operators — are your brand in motion. Every micro-decision they make ripples into customer satisfaction, throughput, quality, safety — and, ultimately, your financials. Build a high-trust, well-tooled environment around them and the impact is immediate and durable.

Empowered front-liners spot the friction others miss and convert it into smarter processes, faster fixes, new service plays that customers feel instantly. Innovation stops being a headquarters hobby and becomes a company habit. Treat frontline culture like a strategic asset and it turns into a profit engine: higher productivity, lower costs, stickier customers, and a steady pipeline of ground-truth ideas. In volatile markets, that isn't just a competitive edge — it's your enduring advantage.

The biggest culture gaps lurk amongst front liners the very people shaping customer reality. When they feel ignored or under-trained for a promotion:

- Service quality nosedives.

- Innovation stalls at the loading dock.
- AI rollouts collect cobwebs instead of ROI.

The Brutal Cost of a Front-liner Culture Crash

Brace yourself: behind the glossy veneer of many corporations lurks a triple-threat money pit that would make any CFO break into a cold sweat. First comes the **Customer-Service Catastrophe**—a global epidemic of horrible interaction, botched calls, tone-deaf chatbots, and mysteriously vanished refund emails that now guzzles a staggering $3.7 Trillion in revenue every single year, up a jaw-dropping 19% in just twelve months. That's not pocket change; that's an economic sinkhole roughly the size of Germany's GDP disappearing because front-liners put customers on hold.

ANNUAL CORPORATE REVENUE LOSS

$8.8 Trillion

Poor Customer Service
$3.7 Trillion XM institute

Turnover / Absenteeism
$2.8 Trillion SIRM

Failed Digital Projects
$2.3 Trillion galapagos.

Illustration Automation Workz

Next whirls the **Turnover Tornado**, an HR nightmare where disengaged employees bail faster than you can print exit interviews. Absenteeism, churn, and the ensuing scramble for replacements can vaporize up to 22 percent of total payroll, once you tally overtime, and sheer lost productivity. Imagine lighting nearly a quarter of your salary budget on fire — then paying overtime to hose down the ashes.

And if that doesn't jolt you, behold the **Digital Disaster**: sloppy change-management practices that torched an eye-watering $2.3 Trillion in failed transformation projects last year alone. To put that in cosmic perspective, you could bankroll about 100 SpaceX missions to Mars with the cash squandered on half-baked rollouts and "we'll-fix-it-in-post" tech launches.

Add these three disasters together and you don't just have a rounding error — you have a tidal wave capable of sweeping even Fortune 100 giants into financial oblivion. In short, ignore the hidden costs of bad service, high turnover, and bungled digital change, and your balance sheet might as well be scribbled in disappearing ink.

Your Wake-Up Call

Companies that elevate front-liners, hard-wire empathy, and celebrate smart risk-taking don't just survive — they dominate the market gobbling up market share and competitors. Think of culture as your secret energy battery: invisible, but once it's charged, the whole organization rockets from 0 to 60 on the revenue racetrack.

A high-culture-ready company:

- ✅ Communicates transparently and often
- ✅ Encourages exploration, not perfection
- ✅ Welcomes questions and experimentation
- ✅ Trains everyone—not just top tech staff—to understand AI tools
- ✅ Builds diverse teams that represent real-world perspectives
- ✅ Empower front-liners, giving them space to lead and test ideas
- ✅ Talks openly about job transformation—not just automation
- ✅ Prioritizes emotional intelligence just as much as technical skills

Corporate executives are investing in AI tools, but have forgotten to create a workplace where employees trust the tools, trust each other, and trust themselves enough to attempt to utilize the AI/tech tools.

Generative AI isn't like older, rule-following tech. It's *creative*. It analyzes massive patterns, draws insights, and generates new ideas—text, images, strategies, and code. But it also requires something no machine can replicate: **human imagination, curiosity, and collaboration.**

To get the best out of Generative AI, employees must be empowered to:

- Think critically
- Ask creative questions
- Learn new skills
- Challenge old assumptions
- Experiment—even if they might fail

But here's the problem: In many companies, culture is the weakest link. Not because people aren't smart or skilled—but because they're scared.

Fear of Job Loss: The Silent Killer of AI Transformation
Let's be honest: AI sounds scary to a lot of people. The moment you say "**automation**," many workers hear "**elimination**." And they're not wrong to worry — especially when headlines scream about robots taking over jobs. According to Pew Research Center, approximately 52% of employees are worried AI will shrink their job prospects, with a third expecting layoff.

Photo Automation Workz

The best AI tools won't get used if your employees think "AI will replace me", "If I learn AI, I'm training my own replacement" Or "Why bother learning something new if I'll be laid off anyway?"

Companies that ignore employee fear will struggle to implement AI. They'll experience:

❌ Slow rollouts   ❌ Minimal usage   ❌ Distrust in leadership

❌ Talent loss   ❌ Wasted technology investments

This fear shows up as:
Silent-treatment - where employees refuse to share tips with the chatbot so it never learns their tricks.

Slow-walk - where employees "accidentally" miss every AI-training session.

Expertise-hoarding – where experienced staff has encrypted institutional knowledge like Fort Knox and refuses to document their processes and procedures or share with teammates.

In cultures where fear rules, no one experiments. And without experimentation, AI doesn't evolve. Accenture found that firms which scale AI *and* nurture a supportive culture enjoy nearly **triple** the ROI.

Here are case studies of a small, middle market and large enterprise who have rewired their culture to scale AI.

Proof It Works — Real Companies, Real Acceleration, Real Profit

AI-Driven Culture Transformation Case Study:
Bank of America (Large Enterprise)

Bank of America founded in 1904 by Amadeo Pietro Giannini as the Bank of Italy has grown into a global financial powerhouse headquartered in Charlotte, North Carolina. With 213,000 employees spanning 42 U.S. states and over 35 countries, the bank serves 69 million clients, operating 3,700 financial centers and managing a digital customer base of 58 million. In 2018 revenue was $87 billion annually. Historically resilient, Bank of America entered the 2020s facing a new frontier: mastering enterprise AI to outpace fintech disruptors, increase operational efficiency, and future -proof its culture.

Problem
Despite record-breaking revenues, Bank of America faced four converging challenges:

1. **Efficiency vs. Scale** – Servicing millions of customers globally meant operations needed a massive speed boost without sacrificing service quality.

2. **Digital Disruption** – Agile fintech firms were delivering AI-powered services at lightning speed, threatening customer loyalty.

3. **Workforce Bottlenecks** – Employees were bogged down by repetitive tasks, IT tickets, and outdated workflows—time that could be redirected toward value creation.

4. **Cultural Stagnation** – A legacy banking mindset slowed AI adoption, creating trust gaps and resistance to change.

These mirrored national HR pain points: the struggle to retain top talent, upskill at scale, and transform culture—issues costing corporations trillions in lost productivity and failed digital initiatives.

Bank of America's bold leap into the AI era was more like steering a supertanker through a digital hurricane. The first issue was legacy tech integration. There were decades of layered infrastructure—that would not be easily integrated with next-gen AI. What should have been a smooth fusion of man and machine often felt like coaxing an antique radio to stream Spotify.

Then came the iron wall of data compliance and security. As one of the world's most regulated financial institutions, Bank of America couldn't afford even a hairline crack in its governance armor. Every AI initiative had to navigate a labyrinth of legal requirements, privacy protections, and industry watchdogs, ensuring that innovation never came at the expense of customer trust.

But perhaps the fiercest headwind was cultural pushback. Many employees — particularly those in traditional roles — viewed AI not as a tool, but as a looming pink slip. This fear, amplified by headlines about job-stealing robots, threatened to stall adoption before it truly began. The narrative had to shift from *"AI will replace you"* to *"AI will empower you."*

Compounding the challenge was the skills gap. Large swaths of the workforce simply didn't speak the language of AI. From understanding how to craft effective prompts to interpreting algorithmic outputs, too many employees were navigating the new frontier without a map.

Finally, the challenge of global scalability loomed over it all. Any AI tool worthy of Bank of America's brand had to function seamlessly across 213,000 employees, spread across 42 states and multiple countries—each with its own regulatory quirks and operational realities

AI People Solution Implementation

Bank of America's cultural rewiring wasn't a tech upgrade; it was a full-scale reinvention of how one of the world's largest banks works, thinks, and serves.

From 2018 to 2020, Bank of America debuted **Erica®**, the first widely adopted AI banking assistant, quickly embedding it into millions of customers' financial routines and building trust in AI-powered money management. Between 2020 and 2023, the focus shifted inward with **Erica for Employees**, an AI colleague supporting HR, IT, and policy needs, especially for front-line and operations staff—proving that easing employees' daily work drives adoption. From 2023 to 2025, the bank entered a phase of **full enterprise integration**, with generative AI embedded across coding, wealth management, operations, and training, giving every employee—from call center agents to financial advisors—a powerful digital advantage.

Organizational Impact and Results

Bank of America's AI gamble didn't just pay off—it rewrote the rules of enterprise adoption. Bank of America increased employee base 4% from 204,000 to 213,000. More than 90% of the workforce in 2025 uses *Erica for Employees*, a digital assistant that has quietly become as essential to daily operations as email. The impact was immediate and dramatic: Calls to the IT service desk plummeted by 50%, freeing tech teams from the tyranny of password resets and printer glitches. Across departments, productivity surged as repetitive, low-value tasks evaporated, giving employees the time to focus on higher-impact work. The customer side told an equally electrifying story. Over 20 million active users now rely on Erica, generating an astonishing 2.5 billion client interactions since launch. Each exchange became smarter, faster, and more personal — transforming routine banking into tailored financial guidance. That precision and speed didn't just delight customers; it drove measurable business outcomes, from increased retention to a spike in cross-sell opportunities.

And the numbers? They sealed the verdict.

Financial Performance

- **Revenue Growth**:
  Increased 121.1% from $87B in 2018 → $192.4B in 2024

- **Revenue per Employee**:
  Increased 111.8% $426,470 (2018) → $903,286 (2024)

- **ROI**:
  $4B annual AI investment delivering 5:1 returns via revenue growth and cost savings.

Bank of America's AI strategy delivered more than operational efficiency — it produced clear financial wins, proving that when culture embraces technology, the bottom line follows in lockstep.

From the moment Erica first said "hello" to the day generative AI began rewriting the bank's

Our AI-driven virtual financial assistant, Erica®, has surpassed

20 million users

BANK OF AMERICA

operating DNA, Bank of America's journey has been a masterclass in marrying cultural change with technological ambition. It's proof that when you blend bold vision with relentless execution, you can teach even the most traditional institutions to think—and act—like the future.

Proof It Works — Real Companies, Real Acceleration, Real Profit

AI-Driven Culture Transformation Case Study:
Dr. Bronner's (Middle Market)

Photo Dr.Bronner's website

Dr. Bronner's is a family-owned maker of natural personal-care products that has been delighting customers since 1948. Headquartered in Vista, California with 226 employees, the company is known for its ethical sourcing, environmental stewardship, colorful labels and versatile castile soaps that line the shelves of major retailers such as Better Health, Costco, Kroger, Publix, Target, Turnip Truck, Walmart, and Whole Foods.

Problem
Dr. Bronner's CEO desired to expand revenue in diverse markets, but struggled to hire critical positions during the pandemic, as the time-to-hire was 99-days and more. The HR department was stretched to its

limits with COVID-19 testing, staff infections and automation of the soap factory. Marketing and PR executives were concerned about rushing any initiatives that could spark fear, confusion, and even resistance among employees who care deeply about the company's "All-One" social-justice values.

AI People Solution Implementation

Conducted a culture audit to understand the intersection of all issues to help them handle growth. The review moved along two tracks. On the public-facing side, the team checked whether the brand's statements on ethics, fairness, and environmental care matched the idea of using intelligent software.

Inside the company, they examined hiring rules, training programs, performance reviews, and even how data moved through the HR system. Surveys and interviews mapped 34 different thinking styles across management, both executives and middle-managers. Emotional-intelligence tests showed how comfortable leaders felt about Key Performance Indicators (KPIs) and data-driven decision making.

Organizational Impact and **Results**
Automation Workz completed a 220-page culture audit report that recommended initiatives to attract potential employees with diverse thoughts, including the expansion of the employment brand and implementation of an AI- based hiring system.

HR staff expanded 400% from three to twelve, adding roles in talent acquisition, employee relations, and data analytics. Middle-managers welcomed the AI-hiring system. Time-to-hire decreased from 99 days to 30 days. Employee population grew 55% from 226 to 350. Current Employees were promoted into new manager and executive roles, including HR Manager, who was promoted to HR Director, VP, HR and now Dr. Bronner's Corporate Board of Director.

Revenue Growth: Dr. Bronner's revenue 2018 $122 million but grew to $209 million in 2024 representing a 70% increase.

Employee Revenue per employee: $597,143 surpassing Amway and Unilever.

Proof It Works — Real Companies, Real Acceleration, Real Profit

AI-Driven Culture Transformation Case Study:
Xylem, Inc. (Large Enterprise)

CEO Gretchen W. McClain launched Xylem in 2011 in Washington, DC as a spinoff of the water-related businesses of ITT Corporation (origin White Plains, NY), with $3.8 billion 12,500 employees worldwide and 270,000 customers.

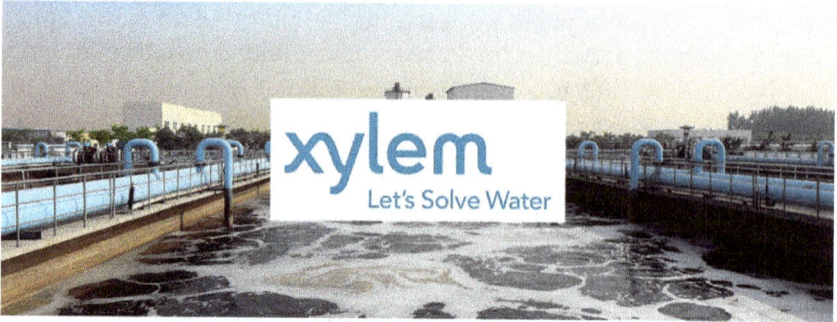
Photo Xylem website

Problem Statement

This water utility was facing critical infrastructure challenges with their aging water distribution system:

- **Aging Infrastructure:** Over 1,000 miles of water mains with an average age of 50 years

- **Increasing Failures:** Water main breaks occurring at an escalating rate

- **Service Disruptions:** Unpredictable outages affecting customer satisfaction and community trust

- **Financial Impact:** Annual pipeline replacement costs reaching $90 million

- **Operational Inefficiency:** Data lived in silos—manual field reports here, GIS maps there, break logs in filing cabinets.

- **Regulatory Pressure:** Need to maintain service standards while managing infrastructure deterioration

Solution and Cultural Fixes

Xylem's first task, therefore, was not purely technical; it was cultural. Company leaders secured visible executive sponsorship at the utility, signaling that an "AI-first" mindset had board-level backing. Cross-functional teams—IT, field operations, engineering, finance—were formed to break down silos and co-design the new predictive-maintenance program. A deliberate pilot approach limited early risk and generated quick, tangible wins that could be broadcast internally.

Parallel to deployment, Xylem funded comprehensive reskilling. Digital-literacy boot camps, mobile-app workshops, and analytics seminars equipped veteran pipefitters and planners alike to interpret AI risk scores and feed better data back into the model. Field operators helped shape a custom mobile app that replaced clipboards with real-time GIS updates — turning skeptics into champions. The tool eliminated duplicate paperwork and reduced midnight emergency calls. Transparent, two-way communication — regular town-hall briefings, dashboards tracking predictive accuracy, and open forums for frontline feedback — maintained trust throughout the rollout. The cultural reset proved as transformative as the technology itself.

Operationally, the utility saw dramatic gains: pipeline failures fell by 75 percent — four times fewer incidents than before — while overall system uptime surged 95 percent. Crews identified and remedied emerging issues twice as quickly, and smarter scheduling translated into a 60 percent jump in productivity.

Workforce Results

- **Digital Confidence:** 90% of employees report increased comfort with technology.

- **Career Development:** 40% of staff received promotions or new responsibilities as data stewards, GIS specialists, and AI super-users.

- **Retention:** 95% employee retention rate during transformation period.

Financial Results

Financially, the AI program proved transformative. Annual pipeline-replacement spending plummeted from $90 million to just $20 million— a 77% cut that generated $70 million in avoided capital outlays. Those savings, combined with lower operating expenses, produced an estimated return on investment exceeding 400% within the first three years. Day-to-day efficiency improved as well, with emergency response costs dropping by roughly 40 percent.

- **Revenue Growth:** 2024 revenues grew to $8.6 billion making its debut on the Fortune 500 list in 2024, ranking #486.

- **Employee Growth:** 23,000 employees worldwide.

Proof It Works — Real Companies, Real Acceleration, Real Profit

AI-Driven Culture Transformation Case Study:
Ford Motor Co. (Large Enterprise)

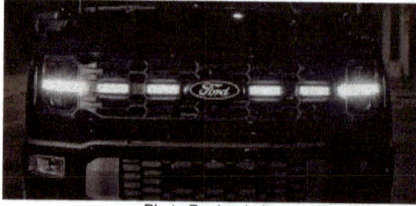
Photo Ford website

Ford Motor Company, born in 1903 under the visionary genius of Henry Ford, revolutionized the world with the moving assembly line and the Model T—transforming the automobile from a luxury toy to a staple of the American Dream. Ford, headquartered in Dearborn, Michigan, operates in over 190 markets, employing 190,000 people worldwide, and commands more than a century of automotive dominance. Today, Ford's race isn't just on the highway—it's in the fast lane of artificial intelligence, electric mobility, and autonomous innovation.

Problem

Manufacturing Inefficiencies – Despite over a century of refining production, human error rates still hovered at a staggering 15–20%, leading to costly rework, wasted materials, and production slowdowns.

Downtime Drain – Unplanned equipment breakdowns chewed through 8–12% of total production time. In the auto industry, that's not just "maintenance trouble"—it's the equivalent of parking millions of dollars' worth of potential revenue on the side of the road.

Supply Chain Snarls – Managing a network of 1,200+ suppliers across six continents with outdated forecasting tools was like trying to drive the Autobahn in first gear. Inventory gluts in one region and shortages in another caused ripple effects that could stall entire vehicle launches.

Front-Liner Skills Gap – Ford's leadership envisioned a sleek, AI-empowered future, but the shop floor told a different story. Many front-line workers had the mechanical expertise of masters, but their AI literacy was miles behind—turning Ford's digital transformation dream into a slow crawl.

Quality Control Blind Spots – Manual inspections, however skilled, were still missing defects that slipped past the factory gates and into customers' driveways. Every overlooked flaw didn't just risk warranty claims—it eroded decades of hard-won brand trust.

Challenge to Implementation

Ford's biggest speed bump on the road to AI mastery isn't buried in microchips—it was lodged in the human mind. The real challenge wasn't coding algorithms or installing robotics it was about downshifting the fears, revving up the confidence, and steering the mindset of an AI resistant global workforce. The resistance came in all lanes.

Veteran Ford employees viewed AI with deep suspicion, fearing it would replace their jobs before proving its value. These anxieties were compounded by outdated training programs built for mechanical skills, not AI-driven operations, and by the need to shift from a reactive culture to proactive, predictive thinking. Integrating AI into decades-old manufacturing systems posed additional risks, as any misstep could halt production. Beyond these internal challenges, Ford also faced complex external hurdles, including navigating global regulatory requirements for autonomous vehicles and safeguarding data from fast-moving cyber threats across continents.

AI Solutions

Ford didn't just implement AI — it unleashed a **four-phase transformation** that blurred the boundaries between grease-stained mechanics and cutting-edge data scientists, turning the factory floor into a high-tech command center.

Ford University AI Platform became the ignition switch for this revolution — a gamified, always-on digital campus where every employee had a personalized AI learning dashboard, cinematic training experiences, and virtual AI coaches guiding them through a fast lane of skill upgrades.

Next came **Cobots & Predictive Maintenance**, the ultimate pit crew of collaborative robots armed with people-aware sensors cycle times with laser precision acting like mechanical fortune tellers, predicting and preventing breakdowns before they could throw production off track.

In **Augmented Reality Training**, workers donned AR headsets that transformed complex repairs into immersive, interactive 3D playbooks—empowering them to make split-second, data-driven decisions with the confidence of a master technician and the insight of a data analyst.

Finally, the **AI Supply Chain Overhaul** reengineered the company's global logistics like a high-speed control tower. Machine learning models fine-tuned demand forecasting, streamlined inventory flows, and optimized delivery routes — all wired directly into Ford's ERP nerve center to keep the entire operation running like a finely tuned engine.

Results so far (2022 to mid-2025)
Ford's AI integration didn't just tweak operations—it supercharged them, delivering results that turned heads across the industry. On the production floor, defects plummeted by **35%** and material waste dropped **28%,** meaning thousands more vehicles now left the line meeting or exceeding quality benchmarks, with fewer costly recalls and a greener environmental footprint. Predictive maintenance became a game-changer, slashing maintenance costs by **31%** while extending the life of critical equipment by **22%,** keeping assembly lines running like finely tuned engines. In the supply chain, smarter forecasting and AI-optimized logistics cut inventory holding costs by **23%** and pushed on-time delivery to a near-flawless **96.8%,** cementing Ford's reputation for reliability and speed.

The cultural shift was just as dramatic. Ford welcomed over 3,000 new hires in advanced tech roles, bringing in AI engineers, robotics experts, and data scientists to fuel the next wave of innovation. In AI-enabled facilities, **78%** of front-line workers reported higher job satisfaction, energized by better tools, more autonomy, and a direct hand in shaping improvements. Innovation surged, with employees generating four times more process improvement ideas—an average of 3.2 per month—and patent applications soaring by an astonishing **89%**. This surge in creativity, technical problem-solving, and cross-functional collaboration signaled that Ford wasn't just adopting AI—it was building a high-performance innovation culture capable of leading the industry into the next era of mobility.

Financial Results
- **Revenue Growth:** Increased 11.1% from Pre-AI (2022) $156.8B from Post-AI (2024): $174.22B

- **Revenue per Employee:** Increased 11.1% from Pre-AI (2022) $825,263 from Post-AI (2024): $917,000

ROI:

- **Investment:** $11.5B
- **Annual Benefits:** $21.62B (revenue increase+cost savings)
- **3-Year ROI:** 188%

Culture is the secret accelerator behind every shiny new tech project. Companies that re-wire their culture earn up to 22% more profit than those that don't bother. The effect is even clearer on the front line: highly engaged teams generate 23% higher profitability and suffer far less turnover. Plug AI into that supportive atmosphere, and the payout surges. Firms that rolled out AI under a people-first culture captured three-to five-times their investment in just 18 months through faster cycles and fewer errors. Employees who feel safe to experiment adopt new AI tools sooner and translate them into business gains more quickly.

Bottom Line: Culture is the best technology investment. Start with a Culture Audit — a high-octane, seven-phase X-ray that exposes your strengths, spotlights your blind spots, and rewires mindsets before the next big tech implementation.

PHASE 1	PHASE 2	PHASE 3	PHASE 4	PHASE 5	PHASE 6	PHASE 7
LEADER DATA	LEADER INTERVIEWS	LEADER SITE	DATA ANALYSIS	COMPILE AND PRODUCE	IN-PERSON ACTION MEETING	LEADERSHIP DEVELOPMENT & COACHING

Illustration Automation Workz website

7-Phases of Culture Audit

☑ **Survey and Collect Leader Data including staff info.**

☑ **Interview Leader and staff to determine viewpoint.**

☑ **Visit Leader Site to understand work environment.**

☑ **Analyze survey, interview and site data.**

☑ **Compile and produce data report with recommendations.**

☑ **Present data report to executives and middle managers.**

✅ **Initiate Leadership Development and coaching.**

Signs You Need a Culture Audit—Right Now

❌ **Staff Disengagement** – Rising Turnover and absenteeism.

❌ **Customer Experience Stagnation** – Customer not satisfied.

❌ **Growth Stall** – AI Spending rises but growth doesn't.

❌ **Front-line frictions** – Barriers to slow or hinder front-line staff.

❌ **Slow handoffs** – Delays in passing work along.

❌ **Silent churn** – Customers leave quietly without notice.

❌ **Shadow processes** – Unofficial procedure workarounds.

❌ **Costly rework** – Waste from fixing errors.

Chapter 3: People Skills - The New AI Literacy

World Economic Forum's Future of Jobs Report 2025 reveals almost **4 in 10 employers** will shrink headcount wherever AI is faster or cheaper, potentially displacing 92 million people worldwide. Although there is a prediction AI will create 170 million new roles resulting in a net increase of 78 million jobs, employees are experiencing

Displacement dread ("Will a bot steal my seat?",)
Skill-shift stress ("Can I learn new tech overnight?!")
Relevance fear ("Does my experience still matter?")

By addressing employees' emotions, anxieties and fears, emotionally smart leaders create a trust cushion, draining the power out of fear, uncertainty, and change to empower experimentation, innovation and growth. In today's AI-powered world, the hottest "coding language" isn't Python or JavaScript—it's **PEOPLE** skills. Unlike old automation tools, generative AI interacts in real time, to generate ideas, solve problems, or write content based on human tone, clarity, and purpose. That means teams must communicate clearly, handle feedback, and collaborate without fear.

People skills are becoming the new AI literacy. The World Economic Forum's *Future of Jobs Report 2025* highlights seven of the top ten "core skills" for 2030 are people skills out-ranking even pure technical know-how. LinkedIn's *Workplace Learning Report 2025* tells the same story from the talent-development trenches: a full 71 % of companies put leadership training at the center of their AI-upskilling strategy, while peer-learning groups, coaching, and internal-mobility programs round out the most effective tactics. McKinsey's July 2025 brief, *We're All Techies Now*, adds the business case: firms that pair advanced digital tools with broad-based "digital fluency **and** human-centric skills" outperform laggards by **two-to-six** times in total shareholder return.

Trust Cushion: The Secret Superpower of Every Fearless Team
Imagine a workplace where employees blurt out a wild idea, admit they messed up, or ask a "dumb" question—and instead of eye-rolls, managers nod with genuine curiosity. That electric sense of freedom is called *psychological safety*. Coined by Harvard professor Amy Edmondson, it is the shared belief that you won't be shamed or punished for taking interpersonal risks. In short, it's the invisible shield

that turns ordinary groups into innovation powerhouses. Why does this matter now? Because artificial intelligence is rewiring jobs at warp speed. When algorithms change daily tasks overnight, teams must learn quickly, experiment boldly, and call out glitches before they explode. None of that happens if people feel judged or silenced. Psychological safety keeps the brain's fear circuit from hijacking performance, allowing creativity and clear thinking to flourish. Think of people skills as the operating system that empowers the digital tools to run smoothly. Self-awareness prevents "automation panic" from hijacking decisions; empathy helps designers build AI that customers actually trust; and social skill turns cross-functional chaos into focused innovation.

How "Emotionally Smart" Leadership Flips Fear into Fuel

✅ **They Read the Room Before the Numbers**
High-EQ leaders notice tense shoulders, quiet Zoom squares, or sarcastic jokes long before turnover stats hit HR dashboards. By naming those emotions —"I sense we're nervous about this AI roll-out"— they drain the anxiety's power.

✅ **They Turn Uncertainty into a Shared Puzzle**
Instead of pretending to know everything, they admit, "We're charting new territory together." That honesty shifts attention from *Who's to blame?* to *How can we figure this out?*

✅ **They Reframe Change as Growth**
Fear whispers, "You could fail." Emotionally smart leaders counter with, "Here's a safe sandbox to experiment." Curiosity blooms when people feel they can test ideas without punishment.

✅ **They Model Vulnerability**
When the manager and executives say, "I'm still learning this tool," it signals that not knowing is acceptable. That single confession makes it psychologically safe for everyone else to ask questions rather than fake competence.

✅ **They Celebrate Questions, Not Just Answers**
Rewarding someone for spotting a flaw or proposing a wild idea teaches the team that exploration beats perfection. The culture shifts from *Hide mistakes* to *Mine them for insight.*

✅ **They Build Trust Rituals**
"One-minute fear dump" at the start of meetings
Anonymous question boards
Post-mortems that hunt for lessons, not culprits
These habits hard-wire psychological safety into daily workflows.

Stuck in the Middle: How Middle Managers Jam the Corporate Engine

Middle managers should be the turbochargers of every company—but too often they're the traffic jam. Why? A critical shortage of people skills. Here's the drama behind the bottleneck:

❌ **"Congrats, You're Boss—Now Forget Feelings!"**
Highly competent individual analytical superstar performers are promoted. Executives assume their charisma is enough to manage people to execute through stress, pressure and crisis.

The result: managers who can individually perform, but stalls when tension rises amongst their teams. Tensions are guaranteed to rise as customers, after the pandemic, have become quite demanding difficult and often offensive.

❌ **Perpetual Squeeze Play**
Orders rain from above, complaints erupt from below, and cross-team pings never stop. With 40 % of their week lost in firefighting, middle managers don't have time for empathy, coaching or self-directed managerial learning. Command-and-control becomes their survival mode — innovation suffocates.

❌ **Motivation Evaporates on the Way Down**
Senior execs may preach "inclusive motivated culture," but without real coaching, that warmth fizzles by two layers down. Strategy stalls in a swamp of status meetings and mixed messages.

❌ **Psychological Safety? More Like Psychological Traffic Cone**
Middle managers avoid hard conversations. Critical risks stay hidden; bold ideas die in silence.

❌ **Innovation Hits the Brakes**
Units led by People-savvy managers adopt new tech 37% faster. The lack of people skills makes every transformation feels like driving with the parking brake on.

Middle managers must be trained as precision bridges, translating executive vision into front-liner action and relaying ground-level insights back to leadership — the key to keeping the corporate race car on track.

Snapshot of the *AI-Ready* Middle Manager

Competency Cluster	Sample Micro-Skills & Tools	Engagement Payoff
Leadership & Influence	Coaching conversations, inclusive decision-making	Clear direction speeds AI adoption cycles
Communication	Storytelling with data, cross-functional translation	Reduces "lost in tech jargon" friction
Creativity	Vision-boarding sprints, lateral ideation games	Fuels innovative AI use-cases
Problem-Solving	Scenario puzzle-gaming, root-cause mapping	Accelerates error-detection/iteration loops
Emotional Intelligence	Personality + 360-EQ feedback	Heightens trust, psychological safety

Here's how to prepare AI-Ready Middle Managers.

☑ **Expose the Blind Spots**: Say goodbye to outdated "soft-skills" webinars. It's time for 360-degree assessments of middle-manager's WIIFM (What's In It For Me philosophy). Employees are more likely to be motivated and committed when they see a clear benefit for themselves.

- Commission middle managers to create and share their personal and professional vision board.

- Determine how much pressure and stress they handle with a personality exam.
- Engage them with interactive puzzle games to understand their technical, leadership, communication, creativity, problem-solving, and emotional intelligence skills/subskills.

- Create their managerial development journey that increases engagement and ultimately drives performance

☑ **Peer-Coaching Pit Stops**: Create small, powerful circles where managers practice tough conversations and share their own leadership mishaps.

☑ **Shadow the Pros**: Let middle-managers learn from the best by pairing them with higher-level executives to job shadow once a month so they can see them diffuse conflicts in real time.

☑ **Pay for People Power**: Transform your compensation structure. Don't just tie bonuses to quarterly profit numbers — align them with people score improvements and cross-team collaboration. Middle-managers are entrusted to motivate front-liners to execute corporate strategies and initiatives. Reward them for motivation of others as it drives profits.

Proof It Works — Real Companies, Real Acceleration, Real Profit

AI-Driven People Skills Transformation Case Study:
Sports Basement (Middle- Market)

Photo Sports Basement Website

Sports Basement opened in 1998, started by Eric Prosnitz and friends in San Francisco with the vision of offering outdoor gear at the lowest possible prices. The company emphasized a local community vibe and accessible shopping experience.

- **2023 Revenue**: Reached $53.2 million
- **2023 Employees**: 394 employees worldwide, 10+ locations

Problem Statement

As sales and community engagement skyrocketed, so did the surge in customer service inquiries. Middle managers quickly noticed a glaring problem— employees spent lots of time crafting responses. This led to slower reply rates, burnout, and a communications nightmare. The goal: cut down customer service email time response by 30-35%, hasten responses, and maintain personal touch.

Challenge to Implementation
"At first, staff met the AI rollout with resistance, fearing it would render their roles obsolete or strip away the personal touch that made customer interactions meaningful. Middle managers found themselves walking a tightrope, having to juggle the technical rollout with addressing deep-rooted concerns and reassuring employees that technology wouldn't replace them, but rather empower their roles."

Solution Implementation
Sports Basement utilized AI to solve their customer service issue. Here is their solution.

- Middle managers led interactive training sessions, pairing newer employees with tech-savvy team members.

- Created a feedback loop: weekly debriefs on how AI was used, spotlighting creative or sensitive cases where human oversight mattered.

- Positioned AI as a collaborative assistant—not a replacement—emphasizing that the highest-rated customer responses were those customized with empathy and context beyond AI defaults.

- Management themselves used Gemini for their own workflow, setting the cultural tone.

Organizational Impact and Results
Employees Reduced Response Time: 30-35%
Increased Employee Count: 495
Revenue Increase: Sports Basement Revenue increased by 11.28% to $59.2 million in 2024.

The success of their AI-driven customer service relied on people skills. Middle managers coached teams on communication, ensuring they showed empathy, tailored to customer. Emotional intelligence is crucial, as staff use AI drafts, but maintain human review to personalize responses. Leaders shifted from micromanaging to focusing on strategic improvements, building trust through transparency and hands-on use of AI. Creativity and problem-solving were encouraged, with managers refining AI-driven processes and using insights from AI to proactively address recurring customer issues. Ultimately, the balance of AI technology and people skills drove success in customer service.

Proof It Works — Real Companies, Real Acceleration, Real Profit

AI-Driven People Skills Transformation Case Study:
Edge Logistics (Middle Market)

Photo Edge Logistics Website

Edge Logistics, founded in 2015 by Michael Rodriguez, a seasoned supply chain strategist and former Fortune 500 VP of Logistics, began as a traditional freight brokerage with a simple mission—connect shippers with reliable carriers, offer transparent pricing, and provide real-time visibility. Operating out of Chicago and Phoenix, the company built a loyal base of mid-market manufacturers and retailers across the Midwest.

By 2022, however, the landscape had shifted. Annual revenue stood at $25.2M with 147 employees (revenue per employee: $171,428), but the freight market was disrupted by venture-backed, AI-native competitors offering lightning-fast, data-driven services. Edge faced a clear choice: adapt or fade into irrelevance.

Problem

Edge Logistics found itself squeezed in a vice grip between the grinding weight of operational inefficiency and the relentless momentum of a transforming market. On one side, manual freight matching was chewing up 45 minutes per load, while customer quote requests languished for an average of 3.2 hours. The lack of consistent, real-time carrier visibility left operations playing catch-up rather than setting the pace. On the other side, profit margins were slipping through their fingers. Reactive pricing—based on yesterday's data instead of tomorrow's market signals—chipped away at profitability each deal. Adding to the pressure, digital-native freight platforms were charging ahead, winning market share with instant quoting and automated matching that made traditional processes look slow and dated.

Inside the company walls, the strain was equally severe. Turnover had hit 33% among coordinators, robbing the business of experience and stability. Recruiting tech-savvy talent for traditional brokerage roles proved an uphill battle, while entrenched departmental silos throttled collaboration.

It was a dangerous convergence—operational drag pulling them backward while the market accelerated forward. Without decisive transformation, Edge Logistics faced not just the erosion of market share, but the risk of fading into irrelevance altogether.

Challenge to Implementation

Edge's AI transformation was never going to be a straight sprint to the finish line. From the start, the company collided with a trio of formidable, almost archetypal barriers. The first was **technical legacy**—an aging transportation management system with limited API capabilities, data scattered like puzzle pieces across CRM, accounting, and dispatch platforms, and a cloud infrastructure so thin it could barely support modern AI workloads.

Then came **human resistance**, a deeply human challenge that no algorithm could simply code away. Middle managers clung to familiar processes, wary of losing control to automation. Coordinators bristled at the shadow of job displacement, while at the top, AI literacy was practically a foreign language among senior leaders.

Finally, there was **strategic hesitation**. Even the most promising AI tools came with a price tag of $20K–$40K annually, and the ROI felt more like a question mark than a guarantee. Integration costs loomed large, and the need for substantial employee training only added to the financial uncertainty.

Together, these obstacles formed a gauntlet that Edge had to run—not just to implement AI, but to rewire its culture and strategy for the future.

AI People Solution Implementation
Edge Logistics made a daring pivot—**people first, technology second**—anchored in the belief that AI succeeds only when humans are empowered to use it.

Key Initiatives:
1. **Leadership & Influence** – "AI Champions" program trained 12 middle managers in AI literacy, turning them into mentors for 10–15 employees each.

2. **Communication Excellence** – Weekly AI updates, monthly town halls, and a company podcast (*The Future of Freight*) kept the story alive and transparent.

3. **Creativity & Innovation** – An *Innovation Lab* gave employees 10% of their time to experiment with AI, producing tools like a predictive load matching algorithm, carrier performance scoring, and dynamic pricing optimization.

4. **Problem-Solving Discipline** – Adopted the DMAIC (Define, Measure, Analyze, Improve, Control) framework to measure, improve, and sustain AI-driven efficiencies.

5. **Emotional Intelligence** – EQ assessments, empathy mapping, and resilience coaching addressed fear and built trust in AI as a collaborator, not a replacement.

Organizational Impact and Results
The results of Edge Logistics' AI transformation were nothing short of seismic. Operational efficiency skyrocketed as once time-consuming processes were slashed to a fraction of their former duration. Load matching, which used to chew up 45 minutes, now wrapped in just 12—an eye-popping 73% speed boost. Quote responses that once dragged on for 3.2 hours were firing back in a lightning-fast 18 minutes, a 91% improvement. Carrier onboarding, formerly a tedious five-day slog, shrank to a brisk two hours, and customer issue resolution times collapsed from 24 hours to just four, redefining what "fast service" meant in the freight world.

The quality gains were equally impressive. On-time deliveries climbed 9%, customer satisfaction scores leapt 19 points to 9.1 out of 10, and carrier retention surged 18 percentage points to an impressive 92%.

Load tracking accuracy hit an almost flawless 99.7%, delivering a level of visibility the industry rarely sees.

But perhaps the most profound shift was in the workforce itself. Turnover dropped from a costly 33% to just 18%, signaling renewed employee commitment. The employee net promoter score (eNPS) swung dramatically 46 points, a testament to rising morale and engagement. Meanwhile, 134 employees earned AI literacy certifications, 42 of them stepped into newly minted roles as "Logistics Intelligence Specialists," 18 account managers became "Customer Success Analysts" with predictive capabilities and 8 employees earned promotions to newly created "AI Operations Manager" position embodying the blend of human skill and machine intelligence that now defines Edge's competitive edge.

Revenue Growth: 65% growth from $25.2M (2022) to $41.7M (2024) —driven by faster operations, better pricing, and higher retention.

Revenue per Employee: 56% increase from $171,428 to $267,308.

ROI: AI investment: $3.2M over 24 months; Cumulative gains: $11.8M **Net ROI:** 268% in two years; **Payback period:** 14 months; **Projected 5-year ROI:** 445%

Edge Logistics didn't just deploy AI—it rewrote its cultural DNA. By prioritizing people skills, before automation, the company leapfrogged into a position as the **"Most Innovative Mid-Market Broker 2024"** in its sector. This transformation proves that AI adoption is not a tech race—it's a human one. When employees believe in the mission, own the tools, and see their value rise alongside machine capability, the return isn't just financial—it's cultural, operational, and unstoppable.

Proof It Works — Real Companies, Real Acceleration, Real Profit

AI-Driven People Skills Transformation Case Study:
Walmart, Inc. (Large Enterprise)

Photo Walmart Website

Founded in 1962 by Sam Walton, Walmart is headquartered in Bentonville, Arkansas, USA, and operates over 11,000 stores worldwide, employing over 2.3 million associates globally. Walmart revolutionized retail with its commitment to low prices, great value and innovative strategies in supply chain management fueling rapid expansion, making it the largest retailer in the world.

Problem:
As Walmart's empire ballooned, executives unveiled their next moonshot: a **digital twin** — a living, breathing virtual clone of every store on the planet. Picture a holographic Walmart where shelf-by-shelf inventory, shopper foot-traffic, and even end-cap layouts update in real time. Fed by armies of sensors and turbo-charged with AI and predictive analytics, this cyber-store lets leaders play out "what-if" scenarios before a single box budges in the real world.

Why it matters
- **Operational Supercharger**: The twin spots empty shelves before shoppers do, rearranges product hotspots for max profit, and slices waste like a chef's knife.

- **Crystal-Ball Forecasting**: It predicts tomorrow's stockouts and next week's trending item, giving managers a head start on fixes.

- **VIP-Level Experiences**: Real-time insights decode customer behavior, serving up hyper-personalized promotions that make shoppers feel like the store reads their minds.

- **Risk-Free Sandboxing**: Need to flip the produce aisle or roll out a flash sale? Test it in the virtual arena first—no forklifts, no overtime, no costly misfires.

Challenge to Implementation
Walmart's digital-twin rollout felt less like a high-tech revolution and more like a tremor rumbling beneath the sales floor. Store Managers — those battle-hardened generals of day-to-day retail warfare—heard the words *virtual replica* and immediately pictured pink slips and powerless clipboards. Would an algorithm now tell them where to move every pallet? Would a glowing dashboard outvote their decades of gut instinct? Their anxiety was turbo-charged by a second jolt: many had never wrangled AI tools in their lives, so the digital twin looked less like a helpful co-pilot and more like an alien cockpit with no instruction manual. Small wonder morale wobbled: the very leaders who kept the

lights on feared they might be dimmed to spectators in their own stores.

AI People Solution Implementation:
Walmart won middle-manager buy-in for its digital-twin AI by pairing hard tech with soft skills:

- **Empathy-Driven Leadership Workshops** Senior leaders kicked off the rollout with frank, highly interactive "Listen & Lead" forums where store managers shared specific fears— from "Will robots run my floor?" to "How do I explain AI to 20-year cashiers?" Facilitators captured pain points on digital whiteboards, then co-created action plans on the spot.

- **Multimodal training** (videos, podcasts, simulators, bilingual cards) matched every learning style.

- **Hack-style labs** Every region hosted a two-day "Digital-Twin Hackathon." Store managers paired with data scientists to drag-and-drop shelves, staff schedules, and promo endcaps inside a live digital twin of their store. Teams ran "what-if" simulations—e.g., What happens to basket size if the endcap for pet treats moves next to seasonal décor?—and then A/B-tested the best ideas in real life. One Florida store boosted same-day pickup efficiency by 18 percent after tweaking its virtual staging before moving a single pallet in the building.

- **EQ Coach Huddles** HR wove EQ into every touchpoint: managers practiced "recognize–reframe–re-energize" conversations to acknowledge associate anxiety, reframe AI as a co-pilot, and spotlight growth paths like becoming an IT coach or leading a store-level AI pilot. Weekly "EQ huddles" let managers swap success stories and vent frustrations in a judgment-free zone, reinforcing psychological safety.

Organizational Impact and Results:
Store Manager Mindset Shift: 87% of middle managers reported feeling "excited" or "optimistic" about AI (up from 38% pre-workshop).

Employee Engagement Score: Increased 9 points

Store Manager job-satisfaction ratings: Highest level in five years.

Trimmed out-of-stocks Inventory: Store Managers co-designed over 120 micro-optimizations inside the digital twin platform to reduce by 14%.

Revenue Growth: $648 billion in revenue in FY 2024, a 6% year-over-year increase,

Conclusion:
Walmart's success with AI implementation highlights the essential role of people skills in driving technological adoption. By focusing on leadership, communication, creativity, problem-solving, and emotional intelligence, Walmart's middle management became the backbone of the company's AI-driven transformation.

Bottom Line: Break the Gridlock
Upgrading the middle managers —specifically through enhanced strategic people tools can transform bottlenecks into smooth highways. When managers master their technical skills, creativity, problem-solving, leadership, and communication abilities, they become a powerful force for change. Better engagement leads directly to better adoption of AI and other transformative tools.

By enhancing middle-manager skills, organizations can break through the gridlock that's holding back company-wide progress to drive better adoption of AI tools and accelerate transformation.

- ✅ **Expose Middle Manager and Front-liner Blind Spots**.

- ✅ **Create Peer-Coaching Pit Stops Group and Solo.**

- ✅ **Shadow the Executive/ Managerial Pros**.

- ✅ **Pay for People Power. Tie Compensation to People Management. Define Metrics beyond Revenue and Profit.**

Chapter 4: Think Like A CEO

Let's get real: the CEO isn't just another executive. The CEO is the brain, heart, and steering wheel of the entire organization. While other leaders operate departments, the CEO operates the *whole machine*. CEOs are the only person constantly juggling vision, risk, money, people, reputation—and the future.

But here's the twist: if a CEO wants to scale, innovate, and survive the AI-powered business storm that's already here...

They need every employee to think like a CEO.
Yes—*even the intern, the forklift operator, the frontline customer rep.*

What Makes a CEO Different Executive

Let's break it down:

Trait	CEO	Other Executives
Vision	✅ Creates, owns & sells it	❌ Executes it within their department
Cross-Function Strategy	✅ Connects all the moving parts like a chess master	❌ Focused on their slice of the pie
Accountability	✅ Answers to the board, investors, press, customers—everyone	❌ One Department unless public-facing
Culture Leader	✅ Sets the tone company-wide	❌ Supports it locally
Spending Power	✅ Moves millions like chess pieces	❌ Manages departmental budgets
Crisis	✅ Front and center when	❌ Manages the mess

Trait	CEO	Other Executives
Manager	things explode	within their zone

The CEO doesn't just **make decisions**—they architect the system behind every decision.

Why a CEO Wants *Everyone* Thinking Like Them

People Make Better Decisions When They See Big Picture "If they don't know the why, they won't care about the how." When employees understand how the business actually makes money, they stop guessing and start leading. They make choices based on what's best for the company, not just their to-do list. Employees can't think like a CEO if they're blind to the company's mission, strategy, or financial model. CEOs must:

- Transparently communicate AI business goals, profit margins, and key challenges.
- Host regular town halls explaining how each role contributes to enterprise growth.
- Cascade KPIs (Key Performance Indicators) to front-liners so they see the direct link between their work and business success.

Imagine a line cook who trims waste to save $10K per year—or a receptionist who fixes a customer frustration no one else sees. That's CEO thinking.

Innovation Explodes When People Feel Like Owners You can't innovate in a box. You innovate when you *own the box.* When employees feel like mini-CEOs, they speak up. They ask, *"What if we tried this?"* They don't wait for permission —they bring solutions.

Most employees don't make strategic decisions because they don't understand how the business works.

A great CEO provides:

- Basic financial training: profit vs. revenue, customer acquisition cost, lifetime value.
- Operational awareness: how different departments connect, how decisions ripple.
- Strategic tools: problem framing, risk/reward thinking, data analysis basics.

This creates what's often called an **intrapreneurial culture**—where people feel like mini-CEOs of their domain. This training also serves as financial literacy to eliminate personal financial stress as they will begin to run their lives like a business.

Some of the world's most disruptive ideas — Toyota's lean manufacturing, Ritz-Carlton's customer-first culture, 3M Post-it ® note—came from front-liners who were trained to think like CEOs.

CEO Mindsets Drive Real Profits When everyone's aligned to the mission, the metrics, and the money, performance skyrockets. According to Gallup: Companies with high employee engagement (AKA "CEO thinkers") outperform peers by **23%** in profitability and **18%** in productivity.

Share trade-offs in decision-making ("We chose X because it gave us long-term stability, not short-term savings.")
Ask "what do you think?" in meetings—even to front-liners.
Pair Front-liners with executive mentors or allow them to job shadow CEO or C-Suite executives.

Employee Base Becomes an Army of Strategic Thinkers AI will replace task-doers and "order takers." But it will empower thinkers. Every company needs people who can spot trends, solve problems, see departments across and think three moves ahead. A CEO can't carry the entire strategy anymore. In fact, they shouldn't. The strategy must live in every cubicle, factory floor, and Zoom call.

If executives only reward output, employees will focus on tasks. If executives reward ownership, insight, and initiative— employees rise to the CEO level. Ways to reward employees is to create "Think Like a CEO" challenges/ pitch competitions where employees live-pitch ideas to cut costs, increase revenue, or improve customer experience. Spotlight staff who identify systemic problems or fix process inefficiencies. Give bonuses or recognition for employees who connect their actions to financial or strategic ideas.

Proof It Works — Real Companies, Real Acceleration, Real Profit

AI-Driven CEO Skills Transformation Case Study:
Joe the Architect (Small Business)

Joe the Architect is a Boston-based boutique firm founded in 2008 by celebrity – bartender – turned – architect Joe Stromer. Twenty-five employees, $8 million annual revenue, mantra: "Make more. Ego less." Designs hot restaurants, sleek offices, luxe homes—powered by people who are "genuinely nice" but always hungry for the next design thrill.

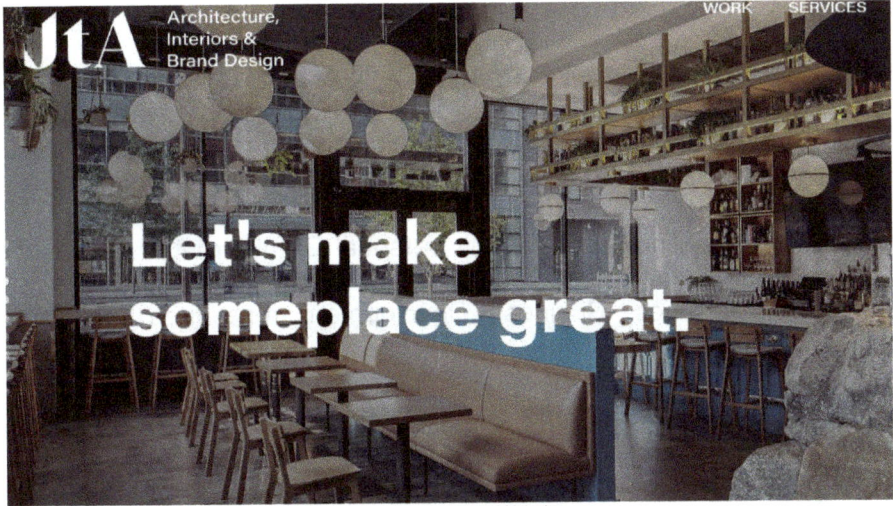

Photo Joe the Architect Website

Problem
Success brought communication chaos: 30-thread email chains, missing change orders, and partners wasting 500+ billable hours a year triaging inboxes instead of drafting skylines. Deadlines slipped, morale sagged, and Boston's architect elite began circling.

Challenge to Implementation
Implementing AI at Joe the Architect wasn't simply a matter of flipping a digital switch; There were four perilous obstacles: **Culture jitters**: designers and project managers feared client conversations would sound canned, draining the very soul from relationships painstakingly built over cappuccinos and site walks.

Workflow risk. Google Gemini couldn't just bolt onto existing Google Workspace tools like an extra app icon; it had to weave itself invisibly into live projects already juggling permits, contractors, and fickle deadlines.

Time poverty. In a boutique studio every hour is billable, so a six-month learning curve felt like a slow bleed no one could afford.

Brand integrity. JtA's calling card—"genuinely nice people"—had to survive the algorithmic makeover. If the new AI engine undermined that signature friendliness, the firm risked gaining efficiency while losing the very chemistry clients came for. Navigating these intertwined challenges required a deployment plan as artful as the buildings JtA designs.

AI People Solution Implementation
- **CEO-Mindset Blitz:** Stromer ordered *every* employee to "Think Like a CEO." Workshops armed juniors and admins with P&L goggles to hunt bottlenecks and ROI gold.

- **DIY AI Playbooks:** Staff built personal Gemini routines—Project Managers synthesized multi-channel feedback, designers auto-generated docs, Business Development reps fired off laser-targeted emails.

- **25 Mini-CEOs, 25 Labs:** Weekly "AI Shark Tanks" where staff pitched and tested micro-experiments, turning Gemini into the firm's multitool.

- **Entrepreneurial Culture Injection:** Everyone asked, "If this were *my* company, how would I unleash AI?"—and then did it.

The AI rollout lit a spark that spread across every corner of Joe the Architect without losing a single team member—empowerment beat burnout, and zero résumés hit the street. Almost overnight, a torrent of ingenuity followed: project managers spun up predictive-timeline dashboards, junior architects unleashed code-compliance bots that caught issues before the city did, and ops staff-built resource-allocation "wizards" that moved people and materials like chess masters.

Clients felt the shift first—e-mails landed faster, proposals read like personal letters, and response times shrank so dramatically that repeat work and bigger retainers rolled in on autopilot. Inside the firm, the numbers told the same story: mind-numbing admin chores shrank by an eye-popping 35 percent, while concept-design cycles accelerated 25 percent, freeing creatives to craft bolder visions and executives to chase the next marquee project.

Revenue Growth: Increased 15–20% capacity boost with $1.2–$1.6 million in new billables
Employee **Revenue per Employee:** Grew 6% from $32 K to $340K
ROI: 850–1,100 % first-year return

Turn every employee into a profit-obsessed mini-CEO, give them AI as a jetpack, even a 25-person studio can roar like a Fortune 500 giant when maneuvered in the right position.

Proof It Works — Real Companies, Real Acceleration, Real Profit

AI-Driven CEO Skills Transformation Case Study: VIPKid (Middle Market)

Cindy Mi, a teenage dropout, founded VIPKid in Beijing in 2013 to connected Chinese students with North American teachers for real-time English lessons to turn screen time into school time.

Problem
By 2017, VIPKid grew to $300 million in revenue then $920 billion in 2019. As demand skyrocketed, VIPKid faced the kind of scaling nightmare most companies only dream of. 100,000 daily lessons. 60,000 teachers. Half a million students. Customer service tickets exploding by 400%. Quality slipping. Margins tightening. That's when Mi did what only the boldest CEOs do—she made every employee think like a CEO.

Challenge to Solution VIPKid embarked on AI high-stakes transformation maze that was failing AI projects were bogged down by bureaucratic approvals and disconnected teams, delaying innovation by up to 60%. Ideas stalled as departments operated in isolation—

crippling speed in a tech landscape that demands agility.

Only 15% of employees understood AI, leaving most staff viewing it as irrelevant or intimidating. Nearly half of VIPKid's educators feared AI would replace them. This automation anxiety caused morale issues and slowed adoption — proving that without trust, even the best tech can't take hold. With over 100,000 daily live video lessons, VIPKid's backend simply couldn't support the massive processing needs of real-time AI.

AI People Solution Implementation
VIPKid knew the real transformation had to start somewhere deeper—inside the minds of their people. So instead of just upgrading tech, they rewired the culture with a radical idea: **what if every employee thought like the CEO?**

Enter the **"Think Like Cindy" program**, a bold cultural experiment that opened the C-suite doors to the entire organization invited employees to sit in on executive strategy meetings, watch high-stakes decision-making in real time, and understand the "why" behind every major AI investment. Suddenly, front-liners weren't just executing plans —they were shaping them.

VIPKid rolled out a **40-hour AI literacy program** for every single employee. Employees didn't just learn *what* AI was — they discovered *how* to use it to solve real problems in their day-to-day work.

VIPKid provided every department head a $50,000 sandbox to run AI pilot projects. No red tape. No approval bottlenecks. Just innovation in action. This wasn't symbolic—it was structural. VIPKid was telling its team: "You don't need permission to lead. Build it."

Organization Impact and Results
AI teaching, content, customer support, and sales systems Improvements were implemented.

Educational Outcomes:
- **Learning Effectiveness:** 28% improvement in standardized English proficiency test scores

- **Student Retention Rate:** Increased from 87% to 95% (industry-leading performance)

- **Lesson Completion Rate:** 23% improvement with optimized teacher-student matching

- **Student Satisfaction:** 40% increase in Net Promoter Score (NPS) reaching 78

- **Teacher Performance:** 25% improvement in teacher evaluation scores

Operational Efficiency: VIPKid didn't just streamline operations—they unleashed a performance revolution. With artificial intelligence at its core, the company reengineered its backend systems to deliver speed, precision, and massive gains in productivity. Customer service, once bogged down by an 18-hour response time, plummeted to just 2.5 hours—a staggering improvement that turned complaints into compliments. Quality skyrocketed too. AI-driven quality control cut lesson complaints by 45%, catching issues in real time and allowing instant adjustments. Students were happier. Parents were impressed. And the company's reputation soared.

Even more impressive with AI handling routine tasks and providing real-time feedback, educators finally felt supported instead of stretched thin. The result? A 40% jump in teacher retention — a major win in an industry plagued by burnout and turnover. VIPKid slashed operational costs per lesson by 35%.

Revenue Growth: $920 Million (2019) to $1.2B (2021)
Customer Acquisition Cost: Dropped from $125 to $69
Customer Lifetime Value: Increased from $850 to $1,148
Revenue per Employee: $136,000 (2017) to $333,000 (2021)
ROI: 607%
Payback Period: 14 months

Celebrate F.A.I.L. - Foster Psychological Safety and Emotional Ownership
People don't think boldly when they're afraid to fail. In most workplaces, failure is something to avoid. It's treated like a mistake, a weakness, or worse—grounds for embarrassment or punishment. But in a workplace that wants to thrive with AI, failure must be reframed. Let's change how we see it.

F.A.I.L. = First Attempt In Learning

Innovation is never perfect the first time. Neither are people. Especially when using something as new and powerful as Generative AI. When companies create space for employees to try, test, and stumble without shame, they spark the exact behavior AI needs to succeed:

🔍 Curiosity   🔧 Experimentation   💬 Feedback

🔁 Iteration   🌱 Growth

Failure is not the opposite of progress. It's the path to progress. As *Amy Edmondson* explains, "The best workplaces aren't the ones with the fewest mistakes—they're the ones where people feel safe admitting them." CEOs must cultivate psychological safety, so employees can:

- Ask questions **"What would I do, with AI, if I were the CEO?"**
- Challenge assumptions
- Share early ideas without judgment

This psychological safety would unleash a workforce, with an inner AI CEO, that acts like owners and creates like innovators — a workforce that masters AI profits.

Proof It Works — Real Companies, Real Acceleration, Real Profit

AI-Driven CEO Skills Transformation Case Study:
Cigna (Large Enterprise)

Photo Cigna Website

The Cigna Group, headquartered in Bloomfield, Connecticut, in 2020, employs 73,000 people who served a staggering 180 million customers across the globe generating $153 billion of revenue. Under CEO David Cordani—an insider since 1991—the company didn't just adapt; it raced to the front of the AI healthcare revolution.

Problem

Cigna was a high-performing ship with a dangerous blind spot — its captains didn't speak the language of AI.

- **Digital Illiteracy at the Top:** Senior leaders couldn't pinpoint high-value AI use cases, let alone assess which tech bets would pay off.

- **Innovation in Silos:** Departments were running their own mini-AI experiments, missing cross-company breakthroughs.

- **Change Paralysis:** Managers couldn't rally teams to embrace AI, leaving new tools underused and underloved.

- **Vision Drift:** Without AI-savvy leadership, divisions ran in different directions—wasting money and momentum.

The result? Slow adoption, wasted investment, and a growing risk of being leapfrogged by bolder competitors.

Challenge to Implementation

The road to AI mastery at Cigna wasn't a straight line of dashboards, data streams, and sleek algorithms—it was a battlefield fought in the minds of its leaders. Technology alone could never deliver the transformation; it demanded a radical rewiring of leadership itself.

First came the **healthcare-specific AI mastery**. In an industry where one flawed algorithm could affect a patient's life, leaders needed more than a casual understanding of machine learning. They required crash courses in AI strategy, ethics, and regulatory compliance, each sharpened to meet healthcare's razor-thin margin for error.

Then there was the **cultural lift for 73,500 people**—an Everest-sized challenge. Rolling out a bold AI vision to tens of thousands of employees was not enough; the real test was making it stick. That required master-level change agents, capable of inspiring, persuading, and sustaining momentum in every corner of the organization.

The transformation demanded **cross-functional orchestration** like never before. AI's true power emerged only when clinical expertise, operational know-how, and technical brilliance converged. Leaders had to smash the silos that had once defined their departments and conduct a symphony of collaboration.

Finally, the leadership playbook for success had to be rewritten. **Measuring what matters** meant looking far beyond cost savings. It meant tracking the synergy between humans and AI, accelerating speed-to-impact, and—most importantly—delivering measurable improvements in patient outcomes.

At Cigna, AI mastery wasn't just about installing smarter tools—it was about forging smarter leaders.

AI People Solution Implementation

Cigna didn't just remove its operational roadblocks—it detonated them with the force of the Leadership Academy, a high-octane fusion of think tank, bootcamp, and AI accelerator. This wasn't another corporate training program; it was a launchpad for a new kind of leader. The mantra was unapologetically bold: **Think like a CEO**—move fast, think big, and wield AI as a force multiplier for every decision and process.

The **Leadership Academy Framework** was designed for velocity and impact.

In **Module 1 – AI Strategy Leadership** (8 weeks), executives dove into ROI modeling, compliance mastery, and visionary mapping for AI opportunities—turning abstract innovation into boardroom-ready business cases.

Module 2 – Creative AI Implementation (12 weeks) unleashed design thinking, advanced prompt engineering, and generative AI lab work, equipping leaders to turn raw technology into market-shifting solutions.

Module 3 – Change Management Excellence (10 weeks) drilled leaders in the art of persuasion, adoption psychology, and hybrid team performance, ensuring that transformation didn't just start—it stuck.

Module 4 – Cross-Functional AI Leadership (6 weeks) trained executives to lead like orchestra conductors, aligning clinical, operational, and technical stakeholders for rapid, unified deployment.

The results were not theoretical—they were tangible and transformative creating new AI products. A Generative AI Virtual Assistant emerged, delivering personalized healthcare guidance to members, crafted by cross-functional teams who blended medical expertise with AI precision. Predictive Analytics for Clinical Profiles took shape, mapping patient risks using Big Data while leaders enforced strict ethical safeguards. Smart Claims Processing, powered by machine learning, obliterated delays, cutting review times dramatically without sacrificing accuracy or compliance.

This wasn't just leadership training. It was the ignition system for a cultural and technological shift that put Cigna miles ahead in the AI healthcare race.

Organizational Impact and Results
Leadership Competency Explosion:
- 95% of senior leaders certified in advanced AI skills.
- Inter-departmental project success rates **up 68%**.
- Innovation ideas from leaders **up 340%**.

Operational Breakthroughs:
- Claims processing time **cut 60%** (14 days to 5.6 days).
- First-call resolution **up 45%**.
- Fraud detection accuracy **up 75%**, preventing $2.1B in fraud annually

Cultural Reset:
- AI adoption in leader-led departments **up 78%**.
- Resistance to change **down 52%**.
- 23 permanent cross-functional AI innovation teams formed.

Clinical Wins:
- Preventive care utilization **up 35%**.
- Hospital readmissions **down 18%**.
- Prescription adherence **up 28%**.

Financial Performance
Customer Growth: Increased 5.56% securing 10 million new
Revenue Growth: 2024 annual revenue is $247.1B an increased of 61.5% from 2020 to 2024

Revenue per employee: Increased to 60.4% to $3,361,904 in 2024 from $2,095,890 in 2020.

ROI:
Total AI + Leadership Academy investment: **$950M** over 3 years.

Annual benefits: **$6.84B**.
Net ROI: **2,060%** (21.6:1 return).

Cigna is doubling down — aiming to graduate 500 new AI-ready leaders annually to pursue more emerging technologies quantum computing for healthcare, and autonomous decision systems in care delivery for health care delivery

With its Leadership Academy, Cigna isn't just riding the AI wave — it's steering it.

Proof It Works — Real Companies, Real Acceleration, Real Profit

AI-Driven CEO Skills Transformation Case Study:
Carewell (Middle Market)

Photo Carewell Website

Carewell, headquartered in Florida, was founded by CEO Bianca Padilla and Jonathan Magolnick after a personal family caregiving crisis to tackle the everyday struggles faced by home caregivers. Carewell is an home health product e-commerce store and mission-driven phone consultation platform designed with the caregiver's experience at its core.

The company quickly gained national recognition, earning a spot on Forbes' list of the *Top 100 Most Customer-Centric Companies* and being named by *Fast Company* as one of the *10 Most Innovative Companies in Retail*. By early 2023, before the introduction of AI, Carewell was already a formidable player in the home healthcare market, generating approximately $45 million in annual revenue with 185 employees.

Problem

The home healthcare market was booming, and Carewell was riding the wave — yet beneath the growth, cracks were starting to show. As demand surged, the company found itself grappling with operational growing pains that threatened to derail its commitment to exceptional service.

The most glaring challenge was customer experience inconsistency. Delivering the same level of empathy, expertise, and efficiency across three major channels — e-commerce, phone consultations, and customer support — proved daunting. The stakes were high: phone consultations alone generated nearly 20% of total revenue, meaning any lapse in service quality risked immediate financial impact.

Behind the scenes, supply chain complexity was compounding the problem. With thousands of SKUs in play, Carewell lacked the real-time demand forecasting needed to prevent costly stockouts and overstock situations. The absence of precision rippled across operations, creating inefficiencies and frustrating both customers and staff.

At the management level, scalability bottlenecks stifled progress. Middle managers—supposed to be the architects of process improvement—were instead spending 70% of their time putting out fires, leaving little room for innovation or long-term strategic thinking.

The company's knowledge management gaps only added to the strain. Product expertise lived in pockets rather than being evenly distributed, meaning customers could get wildly different advice depending on who they spoke to.

Worse still, communication silos kept e-commerce, customer care, and logistics operating like separate islands. Critical issues took longer to resolve, and opportunities for cross-departmental innovation were often missed entirely.

The toll was unmistakable: customer satisfaction stuck at a lukewarm 3.8 out of 5, middle management turnover surging to 28%, average service resolution times dragging at 48 hours, and a troubling 15% stockout rate on essential products. For a brand built on trust and care, the warning signs were impossible to ignore — something had to change, and fast.

Challenge to Implementation

Carewell's leap into AI was anything but a smooth glide — it was a full-on sprint through a maze of complex and costly roadblocks. On the technical front, its BigCommerce platform, while functional for e-commerce, was never designed for the heavy-duty AI. Integrating intelligent systems required extensive customization, all while wrangling fragmented customer data scattered across multiple, disconnected platforms.

But the bigger battle was cultural. Care specialists — prized for their expertise and personal touch — feared AI would dilute their value and make their roles feel transactional. Meanwhile, 65% of middle managers bristled at the thought that algorithms might quietly usurp their decision-making power, shifting authority from human judgment to machine logic.

Then came the regulatory gauntlet. Any AI-generated product recommendations had to pass the high bar of FDA compliance, safeguard sensitive health information under stringent privacy laws, and navigate the thorny question of liability when a machine made the call.

Finally, resource constraints loomed large. The market for healthcare AI expertise was thin, forcing Carewell to weigh the cost of external talent against the daunting task of training existing staff. Budget pressures mounted as leadership juggled the high upfront investment for AI integration with the equally urgent need to fuel rapid business growth.

AI People Solution Implementation

Over 15 intense months, Carewell orchestrated a transformation that blended cutting-edge AI with a relentless commitment to elevating human skill—making sure technology didn't replace people, but amplified their impact. This wasn't just a software upgrade; it was a cultural rewiring with AI as the co-pilot.

It began with **Leadership & Influence Enhancement**. Middle managers were armed with AI-powered dashboards that could predict customer behavior, track team performance, and spotlight market shifts before they hit. But the tech was only half the story—transformational leadership training gave managers the tools to inspire with data, lead change with confidence, and turn insight into action.

The **Communication Revolution** came Next. A natural language processing (NLP) platform became the new secret weapon for customer care — analyzing tone during calls, suggesting the right products in real time, and generating follow-up messages tailored to each caregiver. Behind the scenes, AI automated meeting notes, bridged once-impenetrable departmental silos, and even offered multilingual support. To keep the human touch sharp, employees honed skills in empathetic listening and reading emotional cues.

In **Creativity & Innovation Acceleration**, generative AI scanned caregiver trends to uncover unmet needs, crafting product bundles with precision, and fueled educational content creation. Teams were trained in design thinking, rapid prototyping, and co-creation with caregivers—turning AI insights into human-led innovation at speed.

The **Advanced Problem-Solving Capabilities** phase brought decision-support AI that could offer evidence-based product advice and flag potential service snags before they spiraled. Supply chain teams tapped into AI-driven demand forecasting and automated inventory management, all while learning critical thinking skills and how to make better collaborative decisions in partnership with AI.

Finally, **Emotional Intelligence Amplification** ensured Carewell never lost its heart. AI sentiment analysis didn't just monitor customers—it also kept tabs on employee well-being, spotting burnout risks early. Paired with training in empathy, stress management, and the preservation of the company's family-rooted values, this kept emotional intelligence at the core of every interaction.

Carewell's transformation was not about swapping humans for Machines —it was about creating a workforce that could think, feel, and act at a higher level, with AI as their constant ally.

Organizational Impact and Results

Carewell's AI-powered overhaul catapulted the company into a new league of operational excellence.

On the customer front, satisfaction scores surged from a modest 3.8 to a near-perfect 4.7 out of 5 in just 12 months. Service resolution times collapsed from a sluggish 48 hours to a lightning-fast 8 hours, while phone consultation conversions jumped an impressive 34%, turning conversations into loyal customers.

The supply chain, once plagued by inefficiencies, became a precision engine. Stockouts plummeted from 15% to just 3.2%, forecasting accuracy rocketed by 67%, and product recommendation relevance climbed 45%, ensuring caregivers got the right products at the right time.

For management, AI tools and leadership training drove a 41% boost in middle manager productivity. Engagement scores leapt from 3.4 to 4.5 out of 5, while management turnover nosedived from 28% to a mere 8%, locking in talent that once slipped through the cracks.

In knowledge management, the transformation was just as dramatic. First-contact resolution hit 89%, meaning most customer issues were solved on the spot. Cross-training effectiveness soared by 52%, creating a more agile, multi-skilled workforce ready to tackle any challenge.

Financial Performance
Revenue Growth: Increased 29.6% from 2023 to 2024 with a projected $72.8M for a 61.8% total increase
Revenue per Employee: Increased 22.9% from 2023 $243,243 2024 $298,974
ROI
- AI investment: $8.7M over 15 months
- Net financial benefit (18 months): $19.4M
- **ROI:** 223% with $3.2M in annual cost savings

Carewell rewired middle management into data-driven, emotionally intelligent leaders t5o become a pioneer in AI-enhanced caregiving.

Proof It Works — Real Companies, Real Acceleration, Real Profit

AI-Driven CEO Skills Transformation Case Study:
Caterpillar (Large Enterprise)

Caterpillar Inc., headquartered in Irving-Las Colinas, Texas, was founded in 1925 through the merger of Holt Manufacturing Company and C.L. Best Tractor Company, to become the world's leading manufacturer of construction and mining equipment, diesel and natural gas engines, industrial gas turbines, and diesel-electric locomotives.

Under CEO D. James (Jim) Umpleby III, who took the helm in 2017, it has transformed from digitally driven powerhouse with record profitability and a stock price that surged a 305% increase during his tenure.

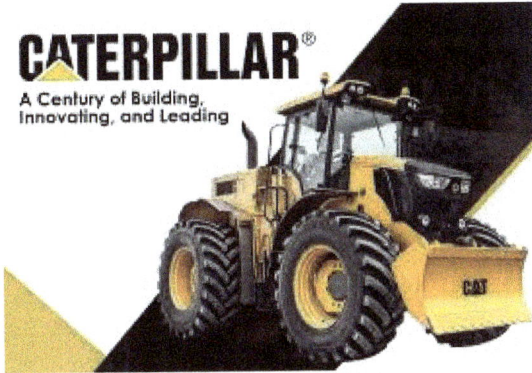

Photo Caterpillar Website

Problem

Before its AI renaissance, Caterpillar was wrestling with a perfect storm of operational headaches that threatened to stall its century-long dominance.

Reactive maintenance meant machines broke down without warning, bleeding time and money as crews scrambled to get them running again. Forecasting was another Achilles' heel—outdated models buckled under the complexity of Caterpillar's sprawling global product lines, leaving leaders to navigate billion-dollar markets with a blurry crystal ball.

Meanwhile, digital innovation was trapped in silos. AI projects lived and died within the IT department, cut off from the rest of the enterprise, robbing the company of the scale and synergy it needed.

Competitors — sleeker, faster, hungrier — were seizing market share with bold innovations, raising the stakes for Caterpillar's survival in an industry shifting under its feet.

And then there was the human factor. Front-line operators, the heartbeat of Caterpillar's operations, were underprepared for the era of smart machines. Without the skills to wield advanced digital tools, even the company's most powerful equipment couldn't reach its full potential. Caterpillar wasn't just facing a tech gap — it was staring down a transformation gap.

Challenge to Implementation

Caterpillar's path to AI mastery was anything but smooth. On the technical front, the company faced the daunting task of fusing cutting-edge AI capabilities with decades-old machinery and legacy systems—industrial giants built for grit and steel, not silicon and algorithms. Data, the lifeblood of AI, came in torrents from across the globe, but it was messy, inconsistent, and scattered across countless formats. Cleaning and standardizing this ocean of information was a herculean effort in itself. And then came the question of scale: how to roll out AI solutions seamlessly across 113,200 employees spread over multiple markets without fracturing operations.

Organizationally, the resistance was real. Engineers and operators, steeped in traditional expertise, eyed AI with suspicion, wary that it might erode the craftsmanship they had spent decades perfecting. Convincing stakeholders to back multi-year AI investments in a fiercely competitive market required not just financial logic, but a bold, unwavering vision. Adding to the challenge was the global talent war—Caterpillar had to recruit elite AI specialists while simultaneously upskilling its massive existing workforce, a balancing act that demanded precision and patience.

Strategically, the stakes were even higher. Executive leadership needed to be in lockstep, and the board had to be convinced that AI wasn't a gamble but a game-changer. Deployment had to be rapid enough to seize market opportunities, yet cautious enough to avoid disrupting critical operations. And all the while, the company had to define clear, irrefutable ROI metrics —proof that AI could deliver tangible value in the unforgiving world of heavy industry. These intertwined challenges set the stage for a sweeping, people-AI implementation that would test Caterpillar's resolve and ultimately redefine its future.

AI People Solution Implementation

Caterpillar's AI transformation was powered by a leadership vision that refused to play small. At the center of this revolution was **Cat Digital**, a bold new arm of the company designed to weave AI into the very fabric of its operations, product development, and customer service. This wasn't a side project or a tech experiment — it was a cultural reset. CTO Otto Breitschwerdt framed AI not as a foreign disruptor but as the next chapter in Caterpillar's century-long innovation legacy, a natural extension of the company's relentless push for progress. Meanwhile, Chief Digital Officer Ogi Redzic made sure AI didn't get trapped in the backrooms of IT. His mission was to break down walls, embedding digital thinking into every department, every process, every decision.

AI was deployed in phases.
Phase One: Foundation Building – Caterpillar didn't just tinker with AI; it built the infrastructure to command it. New digital leadership roles were created, and the Senior Technology Leadership Forum — an elite council of chief engineers and business unit leaders — was assembled to vote with their budgets on where the next wave of AI dollars would go. This put decision-making power into the hands of those who knew the technology's potential best.

Phase Two: Targeted Applications – With the foundation in place, Caterpillar went straight for high-impact wins. Predictive maintenance algorithms began catching equipment issues before they became million-dollar breakdowns. AI-enhanced financial forecasting replaced gut instinct with data-backed precision. Supply chain intelligence platforms made the global movement of parts and products more efficient than ever before, shaving days off delivery timelines and millions off operational costs.

Phase Three: Front-Liner Integration – The most radical move wasn't in the algorithms; it was in the people strategy. Caterpillar paired AI specialists directly with operators in the field through structured mentorship programs. This wasn't top-down training — it was hands-on collaboration, translating technical capability into real-world productivity. Operators, who once feared AI, learned to harness it, becoming champions of the very technology they had once doubted.

And then came the Creative AI Uses that turned heads across the industry. Cat® Construction Technology brought AI-powered operator assistance to heavy machinery, making each job faster, safer, and more precise. Immersive VR/AR/XR experiences revolutionized training, allowing operators to step into virtual job sites, troubleshoot equipment in simulated environments, and master spatial computing tools before setting foot in the field.

Organizational Impact and Results
The operational impact of Caterpillar's AI overhaul was nothing short of a game-changer. Predictive maintenance became the silent guardian of the fleet, spotting trouble before it struck and slashing both downtime and service costs. Machines that once ground to a halt without warning now ran like clockwork, and repair crews swapped panic for precision.

AI-enhanced forecasting turned guesswork into a science, sharpening market predictions and giving Caterpillar the agility to pivot instantly in the face of shifting global demands. On the ground, mentorship programs transformed front-liners from cautious observers into confident AI adopters, ensuring the technology wasn't just implemented but fully embraced where it mattered most.

The ripple effects on innovation were just as electrifying. AI-driven design simulations compressed product development cycles, taking ideas from concept to reality at a speed that left competitors scrambling to keep up. Real-time operational insights reshaped the customer experience, allowing Caterpillar to anticipate needs, personalize solutions, and deliver service with a precision that deepened loyalty and elevated its brand. In short, Caterpillar didn't just make its operations smarter— it made them unstoppable.

Financial Performance
Revenue Growth: Increased 80.6% from 2017 $38.5B to $67.1B in 2023, settling at $64.8B in 2024.
Stock Price: Increased 343% from $94 to $416.54 per share 8/08/25
Revenue per employee: Increased $391,260 to $572,438 (2024)
ROI: surged from $94 to over $381 per share— a $38.5 billion company generated $64.8 billion in 2024 revenue and employs 113,200 people worldwide.

By infusing a Think Like a CEO culture and front-liner mentorship by data scientist, James (Jim) Umpleby III was able to re-wire the Caterpillar culture to turn a century-old industrial giant into an AI leader achieving record profitability, faster innovation, and sustained market leadership.

Now ask yourself—**How would my career catapult if I empowered every employee to "think like a CEO" to increase profits?**

Bottom Line: Every employee should "Think like a CEO." The future belongs to companies where everyone's a strategist, a problem solver, and a value creator. The companies that win won't be the ones with the most executives—they'll be the ones with the most *CEO-minded employees.* Time to groom them.

✅ **Share the Big Picture—Over and Over Again**

- ✅ **Equip Them with Business Literacy**

- ✅ **Showcase "Think like a CEO" with mentoring and/or job shadowing.**

- ✅ **Train every employee to ask - "What would I do, with AI, if I were the CEO?"**

- ✅ **Reward Strategic Thinking, Not Just Execution**

Chapter 5: Creativity Unleashed — Why AI Needs Human Spark

Start Your Engines. Light the Fuse. This isn't your average business transformation. This is the green flag at Daytona. The moment when companies everywhere grip the wheel of AI-fueled innovation, engines roaring.

Welcome to the AI Superhighway where success no longer belongs only to suits and coders, but to every front-liner — superheroes in aprons, hard hats, scrubs, and warehouse boots. These are the new copilots of the digital age. And here's the kicker: the faster AI accelerates, the more it needs human creativity to steer.

AI is your V12 engine. It's the warp drive. But without the *spark* of imagination, empathy, and bold decision-making, that engine chokes on its own smoke. The track gets slippery. The pit crew panics. Game over.

That's why creativity—specifically *front-liner creativity*—is no longer optional. It's the throttle. The steering wheel. The GPS. And right now, too many organizations are treating their people like passengers instead of race-day MVPs.

The Corporate Creativity Gap — Where Good Intentions Stall

Executive teams love to talk creativity. Design-thinking retreats, innovation labs with beanbags and Post-its, cappuccinos and culture decks. Meanwhile, on the shop floor front-liners are handed laminated rulebooks, told to stick to scripts, and trained to follow— but not imagine.

It worked when the road was straight. But today's AI-powered world throws hairpin turns daily. And when automation fails to read emotional tone, or when a chatbot flubs a cultural cue, it's human creativity that can pull the wheel back.

Without it? You stall out. The chatbots crash. The customers ghost. The brand reputation takes a hit. Creativity isn't a luxury anymore—it's the only way to keep from spinning out.

The AI Amplification Effect — When the Stakes Get Loud

Here's what nobody discusses about AI: it magnifies everything. The brilliance of your team? Multiplied. The gaps in communication or confidence? Exposed in 4K resolution. It's like flipping on a jet engine. Every decision made on the floor reverberates across systems. Good prompt? That echo turns into insight. Hesitation? Suddenly, you're five laps behind. When fear replaces curiosity, AI turns into a fragile machine that can't keep up with the race.

The secret weapon? Technological imagination—the ability to see new possibilities inside AI's limits and push beyond them. It's not just about using AI. It's about *surpassing* the manual.

The Human-AI Grand Prix — Who Claims the Podium?

AI can go fast. But only humans make it brilliant. The companies who harness front-liner creativity—who teach their teams to race boldly, to spin out safely, and tune prompts, like precision engines, will dominate. Because at the finish line of this AI revolution, it's not the best tech that wins. **It's the boldest drivers.**

1. Curiosity – The Hidden Radar That Sees What AI Can't

Curiosity is the superpower that keeps your team asking, "What if?" and "Why not?" It's how they spot blind spots, dig deeper, and challenge broken systems before they crash. While AI follows the road—it's curious humans who build new exits.

Curiosity In Action: A warehouse associate notices repetitive packing delays, digs into the data, and uncovers a process bottleneck no one saw—saving hours per shift.

✴ *When curiosity is alive, every glitch becomes a goldmine.*

2. Imagination – The Shortcut No One Else Saw Coming

AI can connect dots. Imaginative humans redraw the map. This superpower is what let's front-liners mix tools, ideas, and intuition to invent fresh solutions on the fly. They don't wait for permission—they create what doesn't exist yet.

Imagination In Action: A customer service rep uses generative AI to not just answer complaints but rewrite the company's return policy into a more empathetic, customer-friendly narrative.

✴ *AI spits out answers. Imagination dares to ask better questions.*

3. Boldness – The Instinct to Hit the Gas When Others Hesitate

Boldness is the fuel that breaks bottlenecks. It's the courage to take the wheel when the playbook isn't working. Bold workers speak up, test wild prompts, and try new moves while others sit frozen at the light.

Boldness In Action: A healthcare technician tweaks a chatbot's triage language without waiting for executive approval—and cuts patient confusion by 30%.

☀ *Boldness breaks the rules—not for chaos, but for progress.*

4. Play – The Launchpad for Low-Stakes Genius

Play isn't wasting time. It's practicing for the main event. When employees are allowed to experiment without fear, they unlock ridiculous insights and powerful ideas. Play is where "what if" turns into "watch this."

Play In Action: A retail team hosts weekly "Prompt Jam Sessions," where employees riff on absurd customer scenarios to find the most creative AI responses. The laughter leads to breakthroughs.

☀ *Play makes innovation fun—and failure forgivable.*

5. Exploration – The GPS Override Your Business Needs

Explorers don't follow instructions—they rewrite them. This superpower sends employees down untested paths, across new platforms, and into AI tools the C-suite hasn't even discovered yet. While others play it safe, explorers pave new lanes.

Exploration In Action: A logistics coordinator tests a beta AI routing tool on side routes and uncovers an overlooked optimization—saving the company $100K in fuel costs.

☀ *They don't just think outside the box—they jump in the rocket and blast off.*

6. Storytelling – The Gear That Turns Metrics Into Movement

Data doesn't move hearts. Stories do. Front-liner storytelling transforms boring dashboards into "aha!" moments that ignite teams and rally support. It's the human voice behind the numbers—and AI just can't fake that.

Storytelling In Action: A line supervisor shares a story of how an AI tool helped a struggling new hire hit record efficiency—and motivates the entire shift team to re-engage with training.

☀ *Storytelling doesn't just report performance. It sells the mission.*

Generative AI is dazzling. But without human creativity, courage, and connection—it's just horsepower with no driver. These six superpowers are the *real* competitive edge in this AI race. If you want transformation that lasts, don't just upgrade your tech. **Upgrade your people.**

Prompt Engineering — The New Driver's Manual

A prompt isn't just a question—it's the ignition key to your AI engine. It's how you tell this hyper-talented, never-sleeps assistant exactly what you want—and watch it deliver at lightning speed.

Think of it like this:
- "Write a funny poem about pizza."
- "Design a logo for a superhero bakery."
- "Explain black holes like I'm 12."
- "Create a business plan for a dog-walking app."

Prompt crafting is high-speed maneuvering on the AI track. Every single word is a turn of the wheel. Every scenario you describe is a lane change that sends AI in a new direction.

The better and sharper your prompt, the better and sharper AI's output. Your frontliners don't need to code— they need to *co-drive* with AI. That means being fluent in how it thinks and knowing when to push it to the edge.

Three Pit-Crew Skills for Prompt Mastery:
- **Language Finesse** – Precision words, smooth phrasing, and sharp analogies pull the best out of AI.

- **Role-Based Prompting** – Setting the scene with "Act as a logistics expert" or "You are my marketing strategist" instantly shifts AI into the right gear.

- **Iterative Testing** – Like tuning a race car: refine, test, tweak, repeat—until your AI laps are clean, fast, and flawless.

Prompt engineering isn't a technical chore—it's the steering wheel for AI strategy. Teach your front-liners how to drive it well, and you'll leave the competition in your digital dust.

How to Fail Workshops — Breaking the Fear-Speed Barrier

Here's the truth: front-liners aren't scared of AI itself. They're scared of the spotlight when things go wrong—the sideways glances from the pit crew, the risk of being "benched" for a misstep, the silent judgment when an idea spins out on the track.

That fear? It's like sand in the engine. It grinds down confidence, slows experimentation, and stops bold ideas from even leaving the garage. And in the high-speed race of AI adoption, hesitation is the real crash risk.

That's where **How to Fail Workshops** come roaring in—deliberately flipping the culture from "don't mess up" to "mess up faster, learn sooner, win bigger." These sessions make failure visible, safe, and *strategic*. Because when failure is part of the training plan, it's no longer a career-ending skid—it's a practice lap.

The Four Power Moves of How to Fail Workshops

- **Crash Simulations** – Teams run deliberately bad prompts to *see* what failure looks like. Not in theory—right there, live. Every skid mark is analyzed for why it happened and how to correct it.

- **Pit-Stop Debriefs** – Quick, no-blame huddles dissect what went wrong and, more importantly, why the risk was worth taking. Every failure becomes a shared learning win.

- **Failure Leaderboards** – Instead of hiding mistakes, these leaderboards celebrate them—ranking the boldest ideas, the most creative misfires, and the risks that showed guts.

- **Turbo-Boost Turnarounds** – Failed ideas are reworked, tuned, and re-released until they run like champions—turning yesterday's breakdown into tomorrow's gold-standard prompt.

Why It Works

Fear shuts down creativity. But give people a safe racetrack for failure, and they'll push the limits of what AI can do. Suddenly, a hesitant "What if...?" becomes a confident "Watch this!" The result?

- **Faster learning cycles** — every lap around the track turns hesitation into instinct.
- **Stronger problem-solving muscles** — your people learn to skid, recover, and still keep their foot on the gas.

- **More innovative AI use cases from the front lines** — not from a whiteboard in the C-suite, but from the asphalt where the real race happens.

When your team isn't afraid to fail, they stop pumping the brakes. They start taking corners faster, testing prompts bolder and thinking bigger. And that's when they stop playing catch-up… and start leading the race.

Proof It Works — Real Companies, Real Acceleration, Real Profits

AI-Driven Creative Skills Transformation Case Study: Carhartt (Middle Market)

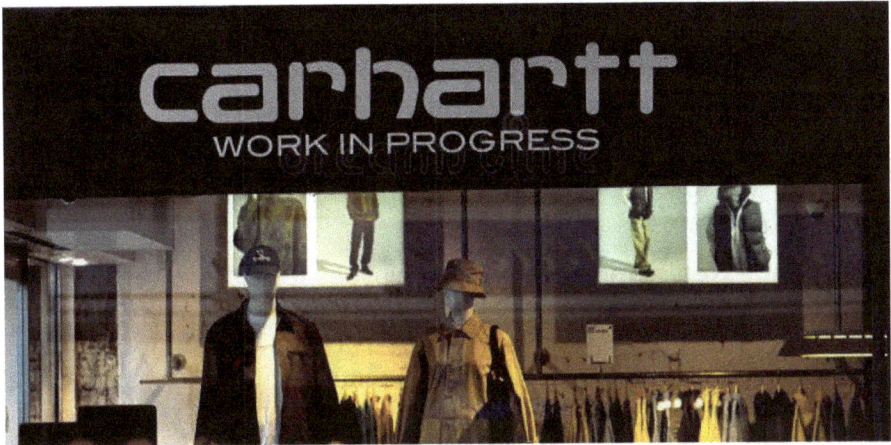

Photo Carhartt Website

Carhartt, Inc.—the rugged titan of American workwear—has been outfitting the backbone of America since 1889. Founded by visionary Hamilton Carhartt in Dearborn, Michigan, the company's DNA is stitched with durability, resilience, and grit. Originally serving railroad workers, Carhartt quickly became the armor of choice for construction crews, farmers, and, more recently, a new generation of style-conscious consumers who value authentic quality.

As of 2023, the company employed 2,006 dedicated workers in Kentucky, Tennessee, and Detroit and generating an estimated $814.7 million in annual revenue — all before its AI revolution began. Still privately held under CEO Mark Valade's leadership, Carhartt proudly remains family-owned and fiercely loyal to its roots, earning a #52 ranking for Best Brands for Social Impact by Forbes in 2025

Problem

By the early 2020s, Carhartt was under siege as E-commerce was rewriting the rules of retail with nimble competitors like Dickies and Patagonia, luring away customers.

Front-liners — retail associates, customer service reps, and inventory managers — were missing the weapons of the modern battlefield: AI-powered tools, real-time data insights, and creative digital problem-solving skills. Without intervention, the risks were stark: loss of profits due to slower response times, lackluster personalization, outdated inventory practices, and costly errors in AI adoption from untrained use.

Challenge to Implementation

Carhartt's leadership knew AI was the future—but getting there wasn't going to be a Sunday stroll:

- **Front-liners lacked AI Literacy:** Many employees had never interacted with AI beyond smartphone voice assistants. Tools like AI-driven chatbots and forecasting engines felt like alien technology.

- **Change Resistance:** Veteran employees feared AI would sideline their expertise, viewing it as a threat rather than an ally.

- **Budget Reality:** As a family-owned company, Carhartt had to make AI adoption lean, cost-effective, and in harmony with its practical, no-nonsense culture.

- **AI Failure Hazards:** Early experiments revealed a darker side—chatbots giving off-the-mark answers, automated tools misreading nuanced requests, and the potential for customer trust to erode if left unchecked.

AI People Solution Implementation

Carhartt didn't just buy software—they rewired their workforce for the AI age, blending human creativity with machine intelligence in a way that was uniquely "Carhartt tough."

1. AI Literacy Training

Partnering with an external AI training provider, Carhartt launched a custom-built front-liner curriculum. Lessons were stripped of jargon, replacing theory with tangible, job-specific applications including:

- Customer service AI that could handle inquiries with empathy and speed.
- Inventory management systems that could forecast like seasoned logistics veterans.
- Prompt engineering workshops teaching staff how to "talk" to AI—crafting precise, creative inputs to unlock powerful outputs, from marketing copy to hyper-personalized product suggestions.

2. Pilot Programs with Feedback Loops
- Chatbots fielded customer questions.
- AI inventory tools predicted demand before shelves ran bare.
- Weekly debriefs with staff became "AI after-action reviews," surfacing failures and reworking prompts until accuracy soared.

3. Cultural Integration
Instead of framing AI as a replacement, leadership positioned it as an exoskeleton for human creativity—taking on repetitive grunt work so employees could focus on the craft of customer care. Low-stakes workshops let staff play, fail, and try again—turning skepticism into curiosity.

4. Continuous Monitoring and Support
A dedicated AI support team stood on call, watching system performance and collecting analytics to fine-tune every interaction, ensuring the AI stayed reliable and human-centric.

Organizational Impact and Results
Carhartt's bold AI transformation paid off with measurable — and remarkable — results:

Customer complaints dropped as AI chatbot accuracy improved by 30%, repeat purchases increased, and stockouts fell by 25% — ensuring that Carhartt shelves were full when demand peaked. Most strikingly, employee satisfaction climbed, with workers reporting that AI had not stripped away their roles but enriched them.

- **Revenue Growth:** Carhartt's online store jumped from **$277M in 2024** to a projected $295–$305M in 2025—a potential 5–10% increase.

- **Revenue per Employee:** Increased **18.6%** Pre-AI, Carhartt earned $406,231 per worker. After AI, that figure rocketed to $481,739.

- **ROI: 620–1,020%.** The $2.5M AI program generated an estimated $18–$28M in extra revenue in year one.

From the sewing machines of 1889 to the AI dashboards of 2025, Carhartt has never been afraid of hard work—or hard change. By arming its front-liners with AI literacy, creative prompt engineering skills, and the wisdom to navigate AI failures, Carhartt didn't just survive a shifting retail landscape—it *thrived*. This is proof that even century-old brands can harness cutting-edge technology without losing their soul.

Proof It Works — Real Companies, Real Acceleration, Real Profits

AI-Driven Creative Skills Transformation Case Study:
IBM Watson Customer Support

Photo IBM Watson Website

IBM, global tech powerhouse headquartered in Armonk, NY with 2023 revenue of a $60.9B has been an AI pioneer for decades —from *Deep Blue* beating Kasparov in 1997 to *Watson* winning *Jeopardy!* in 2011. By 2023, its 33,000-strong customer support force handled millions of interactions annually, generating slightly less in annual renewal and expansion revenue.

Problem
IBM's customer support engine was beginning to lose its competitive edge. Average handle times had crept up to 8.2 minutes, with only 73 percent of issues resolved on the first call. Customer sentiment reflected the strain—Net Promoter Scores stalled at 34, and satisfaction ratings lingered at just 7.2 out of 10. Behind the scenes, support employee turnover reached 18.5% annually, eroding team stability and institutional knowledge. While early AI deployments showed technical accuracy on paper, they faltered in the moments that

mattered most — struggling to read emotional cues, adapt to nuance, or convey empathy. The result was a costly, restless customer base, and a growth path riddled with operational roadblocks that threatened IBM's ability to keep pace.

Challenge to Implementation

IBM's push to fuse AI with human customer service hit a wall of high-speed obstacles.

1. **Technical Grip Issues** – Watson stumbled on emotional language and context-heavy queries accuracy was 67%.

2. **Support Rep Resistance** – 68% of reps feared AI would take their jobs, thus resisted attending training.

3. **Managerial Blind Spots** – Reps lacked the critical skills—prompt engineering, creative problem-solving, and AI collaboration—that could have turned them into co-pilots. Yet, managers dismissed creativity training as unnecessary.

4. **Global Consistency** – 33,000 reps across 170 countries meant cultural, and linguistic complexity.

AI People Solution Implementation

IBM fired up its **Augmented Intelligence for Customer Excellence (AICE)** framework, making AI the *co-driver*, not the replacement:

- **Creativity Training** – 40 hours over 8 weeks, sharpening curiosity, imagination, and boldness.

- **Prompt Engineering Mastery** – Role-based prompting, template libraries, and iterative refinement drills.

- **Failure Recovery Playbook** – Teach reps to spot AI misses and pivot seamlessly to human-led saves.

- **Global Phased Rollout** – $45M over three years, starting with pilots in the U.S., Europe, India, and Brazil.

The results of IBM's human-AI transformation weren't just impressive— they were a full-throttle leap into high-performance territory.

Direct Revenue Impact: Accelerating Growth in Multiple Lanes

IBM's customer support transformation proved to be far more than a service upgrade—it became a revenue-generating powerhouse that accelerated growth across multiple fronts. By reinventing the customer experience and positioning its human-AI collaboration model as a competitive advantage, IBM unlocked new streams of income and fortified its market position.

One of the most striking wins came in customer retention. A measured 8.3 percent increase in retention rates translated into an estimated $340 million in annual revenue that might otherwise have slipped away. This wasn't just keeping customers satisfied—it was securing their loyalty and the long-term value they brought to the company.

The transformation also sparked significant growth in new service revenue. Strengthened, trust-driven relationships led to $89 million in additional service contracts and expansion opportunities each year.

Customers weren't just staying — they were buying more, investing in broader solutions, and deepening their partnership with IBM.

Perhaps most impressively, IBM's success in integrating human creativity with AI precision became a sellable asset in itself. The human-AI collaboration framework was packaged into a consulting offering that now contributes to the company's massive $16.2 billion consulting revenue stream, turning an internal success story into a marketable product.

Finally, superior customer service capabilities emerged as a true sales weapon. In the high-stakes arena of enterprise contracts, IBM's elevated support performance became a differentiator that helped secure wins and renewals valued at approximately $1.2 billion over three years.

ROI
- **Total Investment:** $45M over 3 years
- **Annual Benefits:** $556M (Cost savings + revenue gains)
- **Annual ROI:** 3,607%
- **Payback Period:** 4.1 months
- **Revenue per Employee:** Increased $13,000

In short, this transformation didn't just make IBM's customer service faster and smarter—it turned it into a high-performance revenue engine, proving that exceptional service can be one of the most profitable investments a company can make.

Proof It Works — Real Companies, Real Acceleration, Real Profits

AI-Driven Creative Skills Transformation Case Study:
Zappos (Middle Market)

Photo Zappos Website

Zappos headquartered in Las Vegas, Nevada turned "selling shoes online" into a masterclass in delivering happiness—and did it at scale. In 2020 it became a subsidiary of Amazon with $2.2 billion in annual revenue with an employee force of 2,800. Tony Hsieh's created a culture-first playbook empowering front-liners to wow customers building a brand that made friendliness its competitive superpower.

Problem
By 2022–2023, Zappos' growth started grinding its own gears. Those legendary, heart-on-the-sleeve 45-minute calls that made the brand iconic created wild swings in quality from rep to rep. Meanwhile, rivals weren't just catching up; they were weaponizing automation unleashing 24/7 chatbots that answered in seconds, self-service returns portals that issued labels without human hands, AI triage that routed issues before a person blinked, robot-assisted warehouses that picked orders at machine speed. Rivals layered on dynamic pricing engines, hyper-personalized recommendations, and proactive shipment alerts—driving the cost-to-serve toward pennies creating high expectations for instant service.

Zappos' high-touch model suddenly looked expensive and slow next to frictionless, automated experiences. And to make matters worse, the company's best "wow" moments weren't being captured systemically — brilliant one-offs stayed trapped in individual heads instead of becoming institutional muscle memory. In a world racing to automate, Zappos was paying premium dollars for handcrafted magic—without scaling the recipe.

Challenge to Implementation
Zappos faced an incredible risk of adding AI's jet engine without gutting the soul of Zappos built on human discretion of customer service representatives.

Autonomy vs. Standardization:
Don't cage the reps who make magic.

Authenticity at Scale:
Keep the human spark while adding digital horsepower.

Training Philosophy Clash:
"Break the rules for the customer" meets "follow the model."

Quality vs. Speed:
Efficiency gains can't flatten the brand's personal touch.

AI People Solution Implementation
Zappos didn't roll out AI to replace its people; it harnessed AI to amplify them. The company framed the technology as an Augmented Creativity engine—a tireless co-pilot that feeds context, sparks ideas, and clears busywork—while preserving the final say for human judgment. In other words, algorithms handle the grunt work; humans deliver the goosebumps.

At the heart of the effort was a $12 million, 18-month **Creative Intelligence Program**. First came Curiosity Amplification, where reps learned to pull richer context in seconds and ask sharper, more personal questions. Next, Solution-Generation Workshops recast AI as a brainstorming wingman that could surface ten viable paths instead of one. Finally, Storytelling & Emotional Intelligence training helped teams transform raw insights into moments that actually mattered—little flashes of humanity that customers remember and repeat.

Because Zappos lives and dies on authentic connection, the company engineered **Prompting for Connection**. Reps practiced empathy-driven prompts, used 10× option-generation to widen the creative funnel, and leaned on personalization cues to strike real rapport fast. The goal wasn't scripted niceties; it was faster access to the right human move — the note, the surprise, the fix that feels genuinely thoughtful.

Crucially, Zappos built **Failure→Fuel Protocols** so misfires made the system smarter. If the model missed the moment, reps executed a seamless human takeover, applied a creative override, and logged what worked. Those "human wins" were captured meticulously and looped back to upgrade prompts, playbooks, and policies. Every edge case turned into reusable wisdom.

The rollout was deliberate and disciplined: January 2023 through June 2024, starting with a 300-rep pilot and expanding to all 2,800 reps. Each rep completed 60 hours of upfront training with monthly upskilling thereafter. By the end, AI wasn't a dashboard—they had a culture of co-creation, where technology accelerated context and choice, and humans kept the soul.

Organizational Impact and Results

Zappos' AI-era makeover didn't just feel better—it moved the numbers. On the customer side, satisfaction jumped 4.3%, while recommendation power illustrated by the net promoter score surged 14.1%. Repeat purchases increased 11.0% increase, and customer shout-outs on social media exploded 127%. In short: more joy, more loyalty, more love.

Operations shifted into an higher gear, too. The team handled 23% more conversations without sacrificing quality, while time to fully resolve an issue fell 18%. Frontline ingenuity didn't get lost in the rush; documented creative fixes skyrocketed 156%, and knowledge sharing across teams improved 89%—turning one-off wins into repeatable playbooks.

The people metrics tell the same story. Employee engagement rose from 13.6%, promotions increased 41%, and voluntary turnover dropped 41.7%. Even better, 94% of reps reported feeling more confident and creative, proof that culture matters.

All that momentum fed an innovation flywheel—1,247 process improvements captured. Better experiences, faster operations, stronger teams…and a compounding profitability engine.

Zappos' service magic didn't just warm hearts—it increased annual revenue $234M from repeat customers, larger customer orders and more customer referrals. Revenue per employee climbed 11% from about $786K to 870K per rep.

ROI
- **Investment:** $12M over 18 months; $8M annual maintenance.
- **Annual Benefits:** $264M (rev growth + efficiency + Customer Acquisition Cost reduction).
- **Annual ROI:** ($264M − $8M) / $8M = 3,200%.
- **18-Month ROI:** ($396M − $12M) / $12M = 3,200%.
- **Payback:** 2.1 months.

Zappos proved you can scale humanity with AI. By aiming AI at *creativity, context, and connection* — not just speed — they preserved the brand's soul and supercharged its economics. The result reads like a feel-good blockbuster with a CFO-approved ending: happier customers, energized employees, and a profit engine that just won't quit.

Proof It Works — Real Companies, Real Acceleration, Real Profits

AI-Driven Creative Skills Transformation Case Study:
Verizon Communications Inc. (Large Enterprise)

Photo Verizon Website

Verizon Communications Inc., headquartered in New York City, New York, stands as one of the United States' most formidable forces in telecommunications, wireless services, broadband, and digital solutions, emerged from the historic 1983 breakup of AT&T. and a 2000 merger with GTE. Verizon embraced a technology-first mindset, refusing to be satisfied with merely building networks. Instead, it sought to transform how people, businesses, and devices communicate — laying the groundwork for leadership in the AI-powered era of connectivity. By 2023, before the full integration of AI into its operations, Verizon reported annual revenues of $134.0 billion, employed approximately 105,400 people, and achieved an impressive revenue-per-employee figure of $1.27 million.

Problem

By 2023, Verizon was feeling the heat from every direction. Customer expectations had soared to stratospheric heights, demanding instant service, flawless connectivity, and hyper-personalized experiences. The nationwide 5G rollout, while promising a new era of speed and capacity, was devouring capital at an unprecedented pace. Meanwhile, supply chains — stretching across continents — were snarled in a web of inefficiencies, threatening both delivery schedules and profitability. Inside Verizon's boardrooms, the message was unmistakable: AI was no longer a future consideration; it was the lifeline. Without it, operational excellence would falter, competitive edges would dull, and market leadership could slip away.

The pain points were glaring. Supply chain chaos, fueled by fragmented processes, was causing costly delays and bloated inventories. On the human side, a workforce skills gap loomed large—only 38% of employers nationwide were offering AI training, and Verizon's own front lines weren't prepared for the digital leap. In customer service, legacy systems were throttling both accuracy and personalization, limiting the ability to deliver the kind of real-time, tailored interactions customers craved. And in the network itself, the demands of 5G left no room for hesitation; split-second decisions were essential to keep the infrastructure running at full tilt. For Verizon, the choice was clear— embrace AI as the engine of transformation or risk being left in the wake of faster, smarter competitors.

Challenge to Implementation
Technical Gauntlets:

- Stitching together fragmented data systems into one AI-powered brain.
- Training 105,000+ employees spanning every possible role and tech comfort level.
- Teaching front-liners the dark art of prompt engineering—fast.

Organizational Roadblocks:
- Culture shock for long-tenured employees.
- Balancing AI spend with existing operational demands.
- Overcoming fear that AI might replace, not empower, jobs.

AI People Solution Implementation

Verizon's march into the AI era wasn't a hesitant step—it was an all-out sprint, powered by a training and technology overhaul that touched every corner of the organization. The company knew that to compete at the highest level, it needed more than flashy platforms; it needed an AI-ready workforce, from the front desk to the network control room.

At the heart of this transformation was a comprehensive training strategy designed to build skills at scale. Through a high-profile partnership with edX, Verizon's Skill Forward program unlocked more than 250 courses spanning 84 professional certificate programs. Employees gained fluency in AI fundamentals, mastered prompt engineering, explored the intricacies of machine learning for telecom applications, and sharpened their data analysis skills. This wasn't just reskilling — it was a workforce-wide AI muscle-building program.

But Verizon's talent ambitions didn't stop there. In a powerful show of corporate responsibility, the company teamed up with the Wounded Warrior Project®, pledging to upskill at least 1,000 veterans between Veterans Day 2024 and Veterans Day 2025. These veterans weren't just trained for generic tech roles — they were equipped with cutting-edge AI and digital skills to compete in some of the fastest-growing sectors in the economy.

On the front lines, Verizon targeted one of the most urgent needs in modern business—cybersecurity awareness. Using AI-driven training platforms, employees faced simulated phishing and social engineering attacks, honing their ability to spot and stop increasingly sophisticated threats. These simulations didn't just test knowledge; they built reflexes for a digital battlefield where the stakes are always high.

While people were the centerpiece of the strategy, the technology driving them forward was no less impressive. The OnePlanning AI Platform became Verizon's operational crown jewel, collapsing siloed supply chain data into a single AI-driven command center. This "single source of truth" allowed the company to anticipate disruptions, optimize inventory, and save millions in capital spending—without compromising service or partner relationships.

Customer service also got a major AI infusion. With intelligent assistance at their fingertips, employees could answer 95% of customer inquiries with speed, accuracy, and personalization. The result? A customer experience sharper, faster, and more relevant than ever before.

By weaving together advanced training, veteran empowerment, front-line cyber readiness, and enterprise-grade AI platforms, Verizon didn't just prepare for the future—it positioned itself to define it.

Organizational Impact and Results
Verizon's AI overhaul didn't just modernize operations—it supercharged them. Across the supply chain, AI delivered unprecedented end-to-end visibility, breaking down silos and enabling seamless collaboration with partners. This clarity translated into smarter, faster decisions, keeping products and services flowing without costly delays.

On the customer service front, the transformation was just as dramatic. Armed with AI-powered tools, Verizon representatives achieved an impressive 95% accuracy rate in answering inquiries. Response times dropped, satisfaction scores climbed, and customers began experiencing a level of speed and precision that felt almost effortless.

Sales and marketing teams reaped equally powerful rewards. AI-driven analytics refined lead scoring to pinpoint the most promising prospects, tailored outreach with surgical precision, and delivered recommendations so relevant they boosted both conversions and loyalty. The result was a customer journey that felt less like a transaction and more like a personalized partnership—cementing Verizon's position as a leader in AI-enabled engagement.

Financial Performance
Revenue Growth: Increased 0.6%
Revenue per Employee: Increased 0.79%

Proof It Works — Real Companies, Real Acceleration, Real Profits

AI-Driven Creative Skills Transformation Case Study:
Mayo Clinic (Large Enterprise)

Photo Mayo Clinic Website

Mayo Clinic, founded in 1889 by Dr. William Worrall Mayo and his sons in Rochester, Minnesota, is the world's largest nonprofit integrated health system. Built on the revolutionary principle of integrated practice—where specialists work together to deliver comprehensive care. Mayo Clinic Pre-AI (2019–2020) had 76,000 staff, including 7,300 physicians, Annual Revenue: $13.8B where Revenue per Employee was $181,579 and Patient Volume: 1.3M+ annually. For over a century, Mayo Clinic has led in research, innovation, and patient outcomes—positioning it as a natural

Problem
By 2020, Mayo faced mounting pressures in a changing healthcare environment:
1. **Scalability of Expertise** – Centralized decision-making slowed AI adoption in a high-demand, complex medical environment.

2. **AI Literacy Gap** – Medical training lacked prompt engineering and AI interaction skills.

3. **Administrative Inefficiency** – Manual documentation, scheduling, and capacity management ate into clinical time.

4. **Regulatory Complexity** – FDA approval and compliance requirements slowed AI deployment.

5. **Data Silos** – Clinical and administrative data resided in separate systems, limiting AI's potential.

Challenge to Implementation
Mayo Clinic's push into AI was anything but plug-and-play—the road ahead was paved with formidable challenges that tested both its technical muscle and its cultural resilience.

On the technical front, integration loomed large. Marrying cutting-edge AI tools with Mayo's sprawling, entrenched EHR systems was like rewiring a jet mid-flight. The stakes were high: every glitch could ripple across multiple locations and specialties. Even when systems connected, data quality and standardization remained a constant battle, demanding meticulous alignment to ensure AI models could learn from clean, consistent information. Add to that the immense computational demands of training and deploying sophisticated AI models, and it became clear that raw processing power was just as critical as clinical expertise. And through it all, Mayo had to walk the tightrope of balancing automation with human clinical oversight—because in medicine, trust is earned, not outsourced.

Organizationally, the challenges were equally daunting. Many clinicians feared AI might edge out human judgment, a concern that could stall adoption before it began. The AI enablement team — just 60 people strong—was already stretched thin, struggling to keep pace with the surging demand for generative AI solutions. And then came the Herculean task: training more than 76,000 employees in AI literacy while building governance frameworks that could ensure safety and compliance *without* strangling innovation.

Regulatory and ethical hurdles only added to the gauntlet. Navigating the FDA approval process for AI-powered medical devices was a slow, high-stakes chess match. Ensuring patient data privacy and security in an era of escalating cyber threats demanded vigilance bordering on obsession. Every AI model had to be transparent and explainable to satisfy both regulators and clinicians, while the looming question of liability for AI-assisted diagnoses and treatments hung like a sword over every deployment. This wasn't just AI implementation — it was an all-out campaign to rewrite the rules of modern healthcare without breaking the trust that had taken over a century to build.

AI People Solution Implementation
Mayo Clinic didn't just adopt AI—it reimagined the way an entire health system could innovate. At the heart of this transformation was an enablement -f first strategy that flipped the traditional IT script. Instead of acting as gatekeepers, the data and AI team positioned themselves as *enablers*, arming clinicians with the tools, training, and autonomy to build AI applications themselves. This bold move unleashed a wave of "citizen developers," turning doctors, nurses, and specialists into hands-on AI innovators.

The backbone of this revolution was the AI Factory Platform, built on Google Cloud's Vertex AI. This powerhouse combined a central data library, regulatory compliance tools, and multi-source integration capabilities—creating a one-stop engine for AI development within one of the world's most complex healthcare environments.

Training was equally ambitious. Prompt Engineering Workshops left their mark, with 95% of participants reporting sharper skills and 88% noting improved efficiency with generative AI tools. Data literacy programs began in the C-suite and cascaded down through every layer of the organization. Clinicians had the chance to earn a Medical AI Certification, a specialized in-house credential that bridged their deep medical expertise with cutting-edge AI knowledge. Regulatory education provided essential guidance for navigating FDA approval pathways and compliance standards.

Mayo also embraced a data stewardship model, empowering clinicians and business stakeholders to own and maintain their domain-specific data—ensuring not only quality and usability but also a deeper sense of accountability and engagement in the AI journey.

The rollout was as strategic as it was aggressive:

- **2020:** The Chief Data & Analytics Officer role was established, signaling top-level commitment.
- **2021:** The AI Factory took shape, and pilot projects began.

- **2022:** AI applications moved from concept to scale across clinical and administrative functions.

- **2023–2024:** Generative AI capabilities and advanced prompt engineering training were integrated into the system.

- **2024 and beyond:** Expansion into hypothesis-driven AI models and FDA-cleared devices positioned Mayo as a frontrunner in medical AI innovation.

Mayo Clinic's AI transformation has delivered a seismic shift in how it diagnoses, treats, and innovates—reshaping both its clinical practice and its organizational DNA.

Clinically, the breakthroughs have been nothing short of revolutionary. A proprietary cardiovascular AI algorithm now detects heart pump issues, including low ejection fraction, from a simple 12-

lead ECG—problems once discoverable only through stress tests. This leap forward was validated when the technology earned FDA clearance for market via Mayo's own spinoff, *Anumana*, marking a milestone in AI-driven diagnostics.

Pushing boundaries even further, Mayo adapted its ECG algorithm for Apple Watch integration, bringing advanced cardiac insights into the hands—and onto the wrists—of everyday consumers. In cancer research, the organization broke new ground with *hypothesis-driven AI*, designed to make complex treatment algorithms more interpretable and clinically actionable, opening doors to more personalized and effective therapies.

Operationally, Mayo has achieved performance at scale that would have been unimaginable just a few years ago. During the COVID crisis, machine learning models accurately forecast ICU bed needs, enhancing emergency preparedness. Pediatric care planning saw similar benefits, with AI predicting RSV treatment demand and enabling better coordination at the Children's Center. Across the system, generative AI has taken over time-consuming administrative work—automating forms, clinical note creation, and other documentation—freeing clinicians to focus on patient care.

The transformation has been equally profound in workforce development. More than 1,000 clinicians have been trained in advanced prompt engineering techniques, giving them the skills to collaborate seamlessly with AI systems. A dedicated 60-person AI enablement team now drives deployment and innovation across the organization, while clinical departments have launched their own self-funded analytics teams to control and customize their AI strategies. Executive-led data literacy programs ensure AI fluency starts at the top and cascades across all levels of the enterprise.

Perhaps the most powerful change has been cultural. What began as cautious curiosity has evolved into full-scale enthusiasm, with clinicians shifting from AI skeptics to champions. IT and clinical teams now work side by side, co-creating solutions that are both technically robust and clinically relevant. Data stewardship has reached new heights, improving the quality and usability of information across every department. And, in a true democratization of innovation, Mayo has fostered a "citizen developer" movement — empowering clinicians to design and implement their own AI tools, ensuring that the future of healthcare is being shaped from the front lines.

This is not just an AI adoption story — it is the reinvention of one of the world's most prestigious medical institutions into a faster, smarter, and more collaborative engine for healthcare innovation.

Financial Performance
- **Revenue Growth:** Increased 18.1% in four years 2020: $13.8B to 2024: $16.3B
- **Revenue per Employee:** Increased 18.1% in four years 2020 $181,579 to 2024 $214,474

ROI
Investment: (2020–2024): $150–200M
- **Infrastructure:** $80–100M
- **Training:** $30–40M
- **Personnel & Consulting:** $40–60M

Annual Benefits: $250M
- **Administrative Cost Savings:** $75M
- **Reduced Misdiagnoses:** $50M
- **improved capacity management:** $100M
- **Licensing AI Devices:** $25M

4-Year ROI: 567%
Payback Period: 2 years

In just a few years, Mayo Clinic went from cautious exploration to setting a new benchmark for AI empowerment in healthcare — proving that when you trust your front-liners with the tools of the future, they can rewrite the rules of the game.

Bottom Line: AI delivers its greatest returns when human creativity is in the driver's seat. Technology alone doesn't win the race. The future belongs to organizations that stop treating creativity as a side dish and start serving it as the main course. AI may provide the horsepower, but human spark decides who takes the checkered flag. When organizations unlock front-liner superpowers, AI isn't just a tool—it becomes a multiplier.

✅ **Make AI Training Human-Centric, shared with co-driver.**

✅ **Sanction Play, Experimentation & Failure.**

- ✅ **Reward & Incentivize Curiosity & Imagination.**

- ✅ **Empower Citizen Developers.**

- ✅ **Integrate Creativity with Metrics.**

- ✅ **Create Playbooks to Scale Culture with Technology**

Chapter 6: Storytelling: Road to AI Podium

The world of digital transformation isn't a casual weekend rally — it's a blistering, high-stakes grand prix where organizations roar down the track, gunning for market dominance powered by the raw horsepower of artificial intelligence. Every sector is on the grid, engines revving, with billions in potential prize money on the line. But as the green flag drops, a surprising number of competitors remain stalled in the pit lane, their gleaming AI machines idling in corporate garages — fully capable, but going nowhere.

These stranded giants have invested heavily in the finest algorithms, the slickest dashboards, and the most advanced data pipelines. Yet they're losing ground lap, after lap because they've forgotten a crucial truth: in the AI race, technology alone isn't the vehicle — it's just the engine block. Without the fuel of powerful, consistent, and emotionally charged communication, even the most advanced AI systems sputter. It's like fielding a million-dollar race car without a drop of gasoline in the tank.

The truth is, what separates the race leaders from the pack isn't the size of the engine—it's the chemistry between driver and crew, the trust built in the garage, and the shared vision that fuels every lap. In business, that chemistry comes from leaders who treat communication as the ignition switch and storytelling as the throttle, keeping teams engaged through every curve, crash, and surge toward the podium.

The Podium

In racing, the podium is more than a platform — it's the visible proof of victory. It is where Champions Are Crowned. It's where the drivers stand tall, the champagne sprays, the team colors wave, and the world sees who led the pack from start to finish.

In business, the podium is no different. It's market share gained, customer loyalty earned, and employee pride on full display. It's the quarterly results that silence doubters, the industry awards that validate innovation, and the LinkedIn posts from employees bragging about being part of the winning team.

The journey to the podium doesn't happen by accident. It's the result of precision engineering (your AI technology), the chemistry of a skilled pit crew (your employees), and a race strategy that combines speed with endurance (your leadership). Every metric tracked, every training lap completed, every story told internally contributes to those final seconds when the flag drops and the victory is undeniable.

And here's the real thrill — when your organization reaches the podium, it doesn't just belong to the CEO or the project leads. Every employee who turned a wrench, answered a customer call, closed a sale, or implemented an AI insight is standing there too. The podium is a collective triumph — a moment when technology, people, and leadership have fired on all cylinders to achieve something that competitors can only watch from the track below.

Because in the AI race, it's not enough to finish. **The goal is to finish first — and to make sure your entire team is up there, basking in the win.**

From retail giants using AI to anticipate inventory needs, to banks deploying generative AI chatbots that slash call center wait times, to manufacturing leaders integrating predictive maintenance that saves millions in downtime — the pace is blistering.

Communication is the engine roaring to life. It blasts the facts like exhaust from a dragster: the business battles ahead, the golden opportunities AI can seize, and the concrete ways it will transform operations from sluggish to operational. Communication provides facts, instructions, and status updates. It ensures employees know:

- What AI tool is being implemented
- When and where it's launching
- How to access it
- Whom to contact for help

In AI projects, effective communication creates clarity and reduces confusion. It ensures that no one is left guessing about deadlines, procedures, or responsibilities. But on its own, communication often lacks the motivational spark to shift mindsets.

Storytelling is the rocket fuel injection. It ignites emotional fires, painting vivid pictures of victory celebrations—the "winner's circle" where champagne flows and trophies gleam—while showing each employee exactly how they'll cross that finish line as heroes. Storytelling gives meaning, emotion, and context. It turns an AI initiative from a "management directive" into a shared mission. Storytelling:

- Frames AI as a *tool for empowerment*, not replacement
- Shares success stories from employees and customers
- Connects AI adoption to the company's larger mission and competitive advantage
- Uses metaphors, narratives, and relatable examples to inspire

When emotion is tied to the change—fear reframed into control, hope linked to action, and trust built through transparency—employees are far more likely to *own* the change.

Research proves that employees don't just accept new technology when they understand its purpose — they embrace it with the passion of race car drivers who live for the thrill of victory. For example, boring communication says: "We're deploying an AI-driven inventory system." Electrifying storytelling roars: "This revolutionary system is your lightning-fast pit crew — predicting shortages with psychic precision, turbocharging restocking speeds, and guaranteeing you'll never lose a single sale to empty shelves. You're not just managing inventory— you're commanding a high-tech war machine!"

Conquering the Emotional Crash Course
Digital transformation unleashes a predictable tsunami of human emotions that can either rocket-fuel acceleration or send the entire operation careening into the wall at 200 mph. Fear dominates the opening laps like a nightmare—will AI steal my livelihood? Can I master these alien systems? Will I be left choking on everyone else's exhaust? McKinsey research reveals that companies fully embracing AI report operating margin explosions of 20% or more within three years, yet employee adoption continues to crash and burn these initiatives.

The cure isn't more mind-numbing training manuals or soul-crushing technical manuals—it's narrative adrenaline shot straight to the heart. When employees see themselves as the starring heroes in the AI blockbuster rather than helpless victims of corporate machinery, terror transforms into electric curiosity, and resistance evolves into passionate ownership. Championship storytelling reframes AI from a job-stealing monster into a superhero sidekick, positioning technology as the ultimate co-pilot that handles boring grunt work while empowering humans to focus on high-octane, creative domination.

Watch how different departments come alive when AI narratives are turbo-charged for their specific thrills:

- **Engineering Teams:** "This AI demolishes the boring straightaways so you can absolutely dominate the death-defying hairpin turns of innovation—where your genius makes the difference between first place glory and fifth place shame!"

- **Sales Teams:** "Your AI co-pilot is a crystal ball on steroids, scouting opportunities miles ahead while your competition is still fumbling with their keys!"

- **Customer Service Teams:** "AI feeds you the perfect racing lines—lightning-fast answers, mind-reading personal service—so every customer feels like they're getting VIP treatment in the winner's circle!"

- **Factory Teams:** "This AI is your precision NASCAR pit crew on rocket fuel—monitoring every heartbeat of your machines, predicting catastrophic breakdowns before they even think about happening, and obliterating downtime so your production line purrs like a perfectly tuned race engine. Every flawless part, every impossible deadline conquered is your personal victory lap!"

When the story transforms their contribution from invisible cog to visible champion, AI adoption stops being corporate torture and becomes the mission everyone would die to win.

The Emotional Engine: How Storytelling Annihilates Resistance
The most devastating weapon against transformation resistance isn't boring logic—it's pure emotional dynamite wrapped in heart-stopping narrative. Storytelling is the secret sauce that connects cold knowledge with blazing emotion, helping us make sense of chaos through stories that set our souls on fire. Stories prove to be transformation kryptonite, obliterating resistance by creating comfort zones so wide that change feels like coming home instead of jumping off a cliff.

Change initiatives are organizational earthquakes that shatter established routines, unleashing tsunamis of uncertainty, resistance, and paralyzing fear. But storytelling becomes the ultimate superpower, inspiring and motivating by painting Hollywood-worthy pictures of triumph, glory, and benefits that make everyone's pulse race with anticipation. When leaders share stories that reflect organizational DNA, they don't just reinforce behaviors—they ignite cultural wildfires that guide everyone toward championship behavior.

The neurological impact of storytelling creates measurable brain chemistry changes that transform employee engagement from flatline to full throttle. Championship change narratives require four turbo-charged steps: understanding your story so completely you can describe it while hanging upside down, honoring the past like a racing legend, articulating a mandate for change so compelling it makes people weep with desire, and laying out a path forward so optimistic it makes positive-themed Disney movies look pessimistic. People experiencing organizational change feel emotional hurricanes that usually fuel resistance, but masterful storytelling communication transforms that hurricane into rocket fuel for accepting revolutionary new ways of conquering the world.

The magic erupts when heart-pounding excitement obliterates soul-crushing anxiety. Embracing storytelling means stepping into the driver's seat of leadership, where you guide others through transformation uncertainties like a champion race car driver, building shared understanding and inspiring action that makes hearts race. Rather than presenting change as corporate punishment, storytelling positions transformation as the adventure of a lifetime — a championship race where everyone has a vital role in reaching the podium and spraying champagne.

Fun, explosive narratives serve as emotional antibiotics that annihilate change resistance like a virus. When AI adoption is framed as a blockbuster adventure with clear heroes, death-defying challenges, and mind-blowing victories, employee skepticism transforms into electric curiosity, fear evolves into heart-racing anticipation, and resistance melts like ice cream in a furnace. A championship story — a change narrative that would make Hollywood jealous —communicates the 'what' and 'why' of change with such power that it prevents resistance before it can even think about forming.

Proof It Works — Real Companies, Real Acceleration, Real Profits

AI-Driven Creative Skills Transformation Case Study:
Nike (Large Enterprise)

Nike, Inc., founded in 1964 by Bill Bowerman and Phil Knight as Blue Ribbon Sports, transformed into Nike, Inc. in 1978, named after the Greek goddess of victory. Initially distributing Onitsuka Tiger sneakers, the company shifted to designing its own innovative athletic footwear and apparel, becoming a global leader with its iconic "swoosh" logo.

Headquartered in Beaverton, Oregon, Nike operates worldwide with over 1,000 retail stores and a robust digital presence. Prior to major AI implementation in 2019, Nike reported approximately 73,100 employees and annual revenue of $36.39 billion in fiscal year 2018.

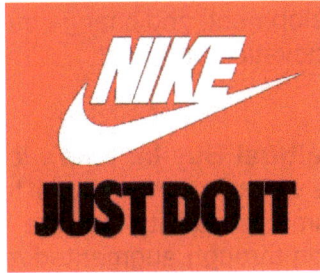

Photo Nike Website

Problem

Nike faced intense pressure to maintain its market dominance amid rising competition from digital-native brands, like Adidas, which leveraged advanced technologies for customer engagement and supply chain efficiency. The company's traditional approaches to product design, inventory management, and customer personalization were becoming outdated, risking slower innovation cycles, stockouts, and reduced customer loyalty in a rapidly digitizing retail landscape.

Challenge to Implementation

Implementing AI across Nike's global operations presented significant challenges:

1. **Cultural Resistance**: Employees, particularly in creative and retail roles, feared AI would undermine human-driven design and customer interactions, leading to skepticism and low adoption rates.

2. **Data Privacy and Ethics**: Handling vast customer data from apps and wearables raised privacy concerns, requiring robust governance to comply with regulations like GDPR.

3. **Integration with Legacy Systems**: Nike's existing IT infrastructure posed compatibility issues for scaling AI tools across supply chains and retail platforms.

4. **Employee Skill Gaps**: Many of Nike's 73,100 employees lacked AI literacy, complicating the deployment of tools like predictive analytics and generative AI for design.

AI People Solution Implementation

Nike accelerated AI adoption by leveraging communication and storytelling, drawing on its brand legacy of inspirational narratives like "Just Do It" to align employees and stakeholders with its digital transformation vision. The methodology combined user-centered design, agile in AI. development, and change management frameworks to foster a culture of innovation and trust.

Detailed Solution:

1. **Storytelling for Cultural Buy-In**: Nike's leadership used internal campaigns, such as videos and town halls, to share success stories of AI enhancing human creativity, like the Nike Fit app improving customer satisfaction through augmented reality (AR) shoe sizing. These narratives positioned AI as an enabler of the "Nike spirit" of innovation, reducing resistance.

2. **Employee Training Programs**: Nike launched AI literacy workshops, training thousands of employees on tools like predictive analytics and generative AI for design. Storytelling was integrated into training, using real-world examples of AI-driven wins, such as optimized inventory reducing waste.

3. **Agile Implementation with Feedback Loops**: Nike piloted AI tools, such as chatbots and demand forecasting algorithms, in select markets, using agile sprints to refine solutions based on employee and customer feedback. This ensured tools were user-friendly and aligned with Nike's brand values.

4. **Ethical AI Governance**: To address privacy concerns, Nike established a governance framework with clear guidelines on data use, ensuring compliance and building trust through transparent communication about AI's role.

5. **Source Citations for Methodologies**:
   The methodology drew from McKinsey's AI adoption framework, emphasizing storytelling and agile development for change management, and Nike's specific approaches as reported in industry analyses.

Organizational Impact and Results

Nike's strategic use of communication and storytelling drove rapid AI adoption, transforming its operations and customer engagement. Key outcomes included:

- **Enhanced Customer Experience**: The Nike Fit app, powered by AI and AR, improved shoe-sizing accuracy, reducing return rates by 20% and boosting customer satisfaction.

- **Supply Chain Optimization**: AI-driven demand forecasting tripled digital order capacity in key markets and reduced stockouts by 25%, improving delivery times.

- **Employee Population Growth**: Population grew from 73,100 to 83,700.

- **Employee Engagement**: Storytelling reduced cultural resistance, with 80% of trained employees reporting confidence in using AI tools, fostering a culture of innovation.

- **Sustainability Gains**: AI-driven material selection supported eco-friendly initiatives, aligning with Nike's sustainability goals and enhancing brand reputation.

Financial Performance
Revenue Growth: Increased 19.4% from $36.40 billion in fiscal 2018 to $51.22 billion in fiscal 2023 with AI-driven digital sales rising from 10% to 26% of total revenue.
Revenue per Employee: Increased 19.4% from Pre-AI: 497,948. Post-AI: $611,947.

ROI:
- **Investment:** $1.5B.
- **Annual Benefits:** $9.6B (rev growth $7.6B + efficiency $2B).
- **5-year ROI:** ($9.6B − $1.5B) / $1.5B = 540%.
- **Payback:** 10.296 months.

The Ultimate Dashboard: Data That Ignites Victory
Traditional corporate dashboards are the equivalent of prison security cameras—tracking every move to enforce compliance, causing employees to slam the brakes instead of flooring the accelerator. The most victorious organizations design dashboards like Formula One command centers on steroids—living, breathing mission control stations that fuel championship performance and celebrate victories like New Year's Eve explosions.

Modern AI capabilities are revolutionizing dashboard functionality with mind-reading powers that would make psychics jealous. AI-powered emotion analytics enable devices to analyze, process, and react to people's moods with supernatural precision while identifying emotional patterns through machine learning that borders on telepathy. Through emotion analytics dashboards that look like something from a sci-fi movie, organizations can centralize their data fortress, providing comprehensive views of customer and employee sentiments and behaviors while showcasing emotional trend patterns over time that reveal the future.

These advanced dashboards serve as emotional crystal balls for transformation initiatives. Emotion AI generates emotion data about workers that reads minds—inferences of emotions, moods, affects, and interior states that provide employers with intelligence that would make the CIA jealous. When employees see declining stress indicators alongside skyrocketing productivity metrics, the dashboard becomes a motivational rocket ship rather than a corporate torture device.

Championship AI adoption dashboards accomplish four earth-shaking functions. First, they celebrate wins by posting "lap times" like victory announcements at the Olympics, showing how AI obliterated processing times or reduced customer response times from glacial hours to lightning-fast minutes. Second, they track the journey using progress bars and real-time leaderboards that make adoption feel like winning the lottery, with each department watching their progress bar rocket toward the finish line like a NASA launch. Third, they link effort to impact with crystal-clear cause-and-effect that would make mathematicians weep, showing how improved performance contributed to business outcomes that shatter records. Fourth, they monitor emotional wellness using AI that detects engagement, frustration, or enthusiasm patterns with such precision that leaders can intervene before problems even know they exist.

Research proves that when employees can see their progress in real time and connect it directly to customer or revenue impact, adoption rates and performance outcomes don't just surge—they explode like volcanic eruptions. The psychology is rocket science simple: people don't just push harder when they feel part of the race — they push with the intensity of Olympic athletes going for gold, not prisoners under surveillance.

Proof It Works — Real Companies, Real Acceleration, Real Profits

AI-Driven Creative Skills Transformation Case Study: Automation Workz

Founded in 2019 by Ida Byrd-Hill in Detroit, Michigan, Automation Workz, a behavioral informatics firm, charged onto the workforce transformation track during the pandemic, with a bold mission: unlock tech career pathways for overlooked front-liners — cashiers, warehouse associates, call center reps — the very people keeping business engines running, but earning only $24,960 to $39,520 a year in or near poverty.

From day one, the company's online learning lab became known for doubling graduate salaries by catapulting them into high-demand certifications in AI & Data Analytics, Cybersecurity Ops, Network Engineering, and IoT Tech Support.

◎ **Individual Program Readiness Snapshot**

Comprehensive overview of user's program readiness and skill development.

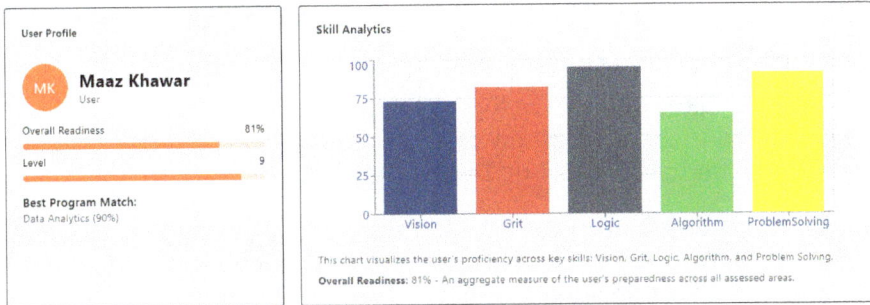

Main Dashboard

User Profile

MK **Maaz Khawar**
User

Overall Readiness   81%

Level   9

Best Program Match:
Data Analytics (90%)

Skill Analytics

[Bar chart showing proficiency levels for: Vision, Grit, Logic, Algorithm, ProblemSolving]

This chart visualizes the user's proficiency across key skills: Vision, Grit, Logic, Algorithm, and Problem Solving.

Overall Readiness: 81% - An aggregate measure of the user's preparedness across all assessed areas.

Photo Automation Workz website

Their early edge came from the Life Culture Audit — an AI-powered behavioral assessment and coaching app that went beyond résumés and job histories to measure raw potential. It used AI to scrape non-traditional data sources — a digital vision board, a personality exam, and a deceptively simple block puzzle game — to generate predictive analysis that exposed hidden tech talent long before traditional HR systems could see it.

From 2021 to 2024, Automation Workz tested the Life Culture Audit with front-liners in the WIOA workforce development system. In 2024, it rebranded and relaunched as SenseiiWyze — a high-octane, AI-powered coaching and analytics platform built to identify, develop, and accelerate internal talent for the AI era.

The Problem
Automation Workz could prove that front-liners could master advanced tech — but scaling that success with employer partners was another

matter. Even with tuition covered by federal workforce development dollars, employers remained blind to the talent already in their ranks.

This wasn't just a skills gap — it was a culture gap. SenseiiWyze could produce hard data showing front-liners' tech readiness, but executives didn't recognize that their own workplace dynamics were stalling talent growth.

At the same time, WIOA-funded programs were enrolling learners into high-tech certifications without assessing academic, emotional, or cognitive readiness. The result? Completion rates in the first year dropped to 7.93%. The real barrier wasn't just training — it was coaching, motivation, and belief.

Challenges to Implementation

1. **No Clarity of Front-Liner Development Pipeline** – Without a culture audit, leaders were steering blind. They had no baseline on engagement, talent bottlenecks, or cultural drag points slowing AI adoption.

2. **Cultural Skepticism** – In the absence of metrics, many HR leaders defaulted to the old narrative: *front-liners aren't technical.* This bias stalled internal promotions before they began.

3. **Data Privacy & Integration** – Scraping and analyzing vision board, personality, and puzzle data raised compliance hurdles with HRIS systems never designed for hidden potential mapping.

4. **Bias & Accuracy in AI** – AI models risked cultural bias without diverse data and ongoing audits.

5. **User Adoption & Resistance** – Leaders feared losing the "human touch" if AI entered the coaching process.

6. **Motivation Gap** – Years of repetitive work had dulled ambition. Employees needed visible career pathways and proof leadership was invested in their growth.

AI People Solution Implementation
The Audit Mindset
Automation Workz learned that the difference between performance and nonperformance often came down to whether leaders measured their culture — and acted on it.

- In companies with no culture audit performed a mini audit of talent development system or a company-wide audit.

- In companies that had run audits but never acted, the issue wasn't insight — it was inertia, staged leadership workshops as "pop-up" audits, mapping engagement gaps, bottlenecks, and hidden talent pools in real time.

In both cases, leaders walked away with undeniable visibility into the friction points throttling productivity, innovation, and AI adoption — and a plan to fix them.

Learner Future Orientation
Front-line workers engaged with the SenseiiWyze mobile application to:
- Select life goals for vision board compilation
- Complete comprehensive Big 5 OCEAN personality assessments
- Progress through at least 10 levels of JavaScript-generating block puzzle games

AI Scraping & Analysis
SenseiiWyze analyzes vision boards (life goals), personality exam results (OCEAN traits), and puzzle outcomes (cognitive flexibility) to reveal an employee's readiness in Technical skills and interpersonal skills - Communication, Creativity, Emotional Intelligence, Leadership, and Problem-Solving.

SenseiiWyze Talent Mapping
Gamified assessments surfaced hidden stars — the forklift driver with cybersecurity instincts, the retail supervisor with a knack for data analytics, the customer service rep who is a prompt engineering guru in her personal life.

Personalized Tech Career Navigation
SenseiiWyze acted as a digital pit crew chief, breaking certification paths into manageable milestones, tying skills to real internal openings, and sending motivational nudges along the way.

Leadership Dashboards
Executives could finally track cultural improvements in real time — seeing how rising engagement connected directly to reduced turnover and faster AI adoption. But the dashboards went further: they delivered targeted recommendations on each learner's most promising tech specialty and showed how their skill levels compared to peers. This allowed leaders to direct precision coaching, mentoring, and stretch assignments that groomed high-potential employees upward into AI, cybersecurity, network engineering, or Tech Support roles.

Tech Skill Coaching with Simulations

Graduates didn't just pass tests — they solved real-world AI, cybersecurity, network engineering and IoT challenges in simulations, proving they could perform under race-day conditions.

Communication & Storytelling Integration

Personalized success stories were shared with executives and staff internally and externally. This storytelling built emotional buy-in from learners and executives alike.

Addressing the Challenges

- **Privacy** – Strict compliance protocols and explicit consent protected user data.
- **Bias Mitigation** – Continuous AI audits and diverse training data reduced skewed results.
- **Adoption** – Workshops framed AI as a "leadership co-pilot," piloted in phases to build comfort.
- **Scalability** – App store partnerships expanded reach beyond pilot programs.

Organizational Impact & Results (36 Months)

Client Talent Development Outcomes:

- Built a dedicated front-liner tech talent pipeline.
- 42% increase in internal promotions to AI, cybersecurity, network, and tech support roles.
- Salaries jumped from $34,000 to $68,000+ post-certification.
- 58% retention improvement, saving $25,000 per avoided turnover.

Client Organizational Transformation:

- Expanded reach to underrepresented groups.
- AI adoption accelerated as trained employees became in-house champions.

Automation Workz Industry Recognition:

- Ranked Top 10 Cybersecurity Bootcamp on Intelligent.com (2022–2023).
- CEO Ida Byrd-Hill joined the BCG/HolonIQ AI & Digital Leadership Council.
- Appearance on the Corporate Training Landscape Map in the IT/Tech Training.

Automation Workz Organizational Transformation:

- Freed Automation Workz staff schedule so instructors could provide high-value mentoring/coaching, boosting Automation Workz productivity.

Financial Performance
Revenue Growth: Increased 340.8% from 2021 to 2024. Projected growth of 2081.8% in 2025.
Revenue per employee: Increased 120.4% from 2021 to 2024
ROI: 762.9%

For companies struggling with AI adoption, the starting line is culture. SenseiiWyze delivers the deep behavioral and cognitive insights that build personalized coaching plans — transforming untapped front-liners into confident, high-demand tech talent.

In the race to become AI-ready, SenseiiWyze isn't just in the fast lane — it's setting the pace.

The Human Pit Stop: Face-to-Face Lightning Strikes
While digital messages can blast through corporate networks faster than light speed, nothing replaces the trust-building nuclear power of face-to-face communication that hits like lightning. Face-to-face interaction isn't just nice—it's the premium rocket fuel that keeps the AI adoption engine roaring at championship levels. When leaders show up in person, they're not just reading the room — they're reading souls through non-verbal cues that make up 55% of communication impact, serving as early-warning radar systems that no dashboard could dream of replicating.

Three high-impact strategies maximize face-to-face nuclear fusion. Town hall kickoffs blast the green flag on AI projects with live Q&A sessions where front-line gladiators voice hopes, questions, and even skepticism that could power small cities. Social learning huddles create cross-departmental dream teams that meet informally to swap victories and AI usage stories that turn adoption into a shared championship quest rather than a lonely death march. Leader roadshows take executives out of ivory tower command centers and onto the battlefield, armed with live AI demos that don't just show — they prove how technology makes work smoother than silk and more rewarding than winning the lottery.

These face-to-face pit stops aren't wasted time—they're critical moments to inspect the racing machine, adjust the championship strategy, and give the team confidence boosts that send them back to work faster and more committed than Navy SEALs on a mission.

Proof It Works — Real Companies, Real Acceleration, Real Profits

AI-Driven Creative Skills Transformation Case Study:
GreenPath Financial Wellness (Middle Market)

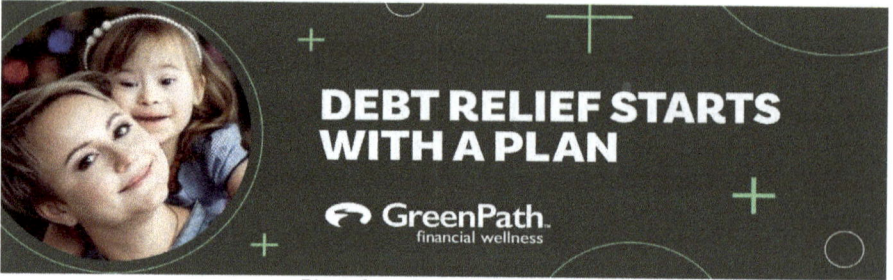

Photo from GreenPath website

GreenPath Financial Wellness, headquartered in Farmington Hills, Michigan, was founded in 1961 by Albert O. Horner as a budgeting and financial education service for the Michigan Credit Union League. It has evolved into an independent nonprofit dedicated to empowering individuals through debt management, housing counseling, and financial education. GreenPath operates nationwide, partnering with credit unions, banks, and employers to deliver judgment-free financial guidance. Prior to AI implementation in 2021, GreenPath had 456 employees and annual revenue of $44,875,928 in 2020.

Problem
GreenPath faced limitations in scaling personalized financial counseling amid rising consumer debt and economic challenges, such as the COVID-19 pandemic. Traditional human-led counseling was resource-intensive, with high costs and limited availability, making it difficult to reach underserved populations 24/7. This resulted in delayed assistance for clients dealing with debt, credit issues, and savings goals, exacerbating financial stress and reducing overall impact.

Challenge to Implementation
Implementing AI for financial coaching presented several hurdles:

1. **Complexity in Personalization**: Ensuring AI could handle diverse financial situations without jargon, while maintaining empathy and accuracy.

2. **Compliance and Regulation**: Financial advice is heavily regulated, risking fines for incorrect guidance and requiring robust monitoring.

3. **Cost and Scalability**: High expenses for expert advice ($200–$500 per session) limited accessibility, and integrating AI required balancing technology with human oversight.

4. **User Adoption and Trust**: Building trust in AI among clients and staff, especially in sensitive financial matters, while addressing potential resistance to technology.

AI People Solution Implementation
GreenPath addressed these challenges by launching the Virtual Financial Coach (VFC) in partnership with eGain in 2021, leveraging AI to deliver 24/7 personalized guidance. Communication and storytelling were central to accelerating implementation, using motivational nudges, client success stories, and behavioral science principles to engage users and staff. The methodology combined AI-driven conversational technology with GreenPath's 60 years of counseling expertise, incorporating gamification and empathetic narratives to foster adoption.

Detailed Solution:
1. **Communication Strategies**: GreenPath used digital channels (text, email, WhatsApp) for seamless, discreet interactions, allowing users to pause and resume conversations. Internal communication involved town halls and training to demystify AI, framing it as an empathetic extension of human counseling.

2. **Storytelling Integration**: The VFC employed bite-sized action steps and celebratory messages (e.g., badges, success narratives) to motivate users, drawing from real client stories like debt payoff journeys shared in podcasts and reports. This created psychological safety and encouraged progress, reducing stigma around financial help.

3. **Addressing Challenges**: For compliance, AI advice was pre-vetted and compliant with regulations. Cost was mitigated by automating routine guidance, freeing human counselors for complex cases. Personalization was achieved through AI personas (e.g., cheerleader style) and A/B-tested nudges.

4. **Implementation Process**: Piloted with 25 partners in 2021, expanded in 2024 with features like credit score analysis and product referrals. Methodologies included user-centered design for

accessibility and behavioral science for nudges, ensuring cultural relevance.

Organizational Impact and Results
The VFC transformed GreenPath's operations by scaling counseling to over 16,837 users in 2022, with 92% valuing 24/7 availability and 84% reporting increased confidence in financial goals.

Storytelling accelerated adoption by humanizing AI, reducing resistance among staff and clients, and fostering a culture of empathy. Results included a 200% growth in downpayment assistance programs, higher client retention, and expanded services to underserved communities. Employee count increased 71 persons. AI freed counselors for high-touch cases, lowering turnover to historic lows and thus earning 2023 Top Workplaces awards.

Financial Performance
Revenue Growth: Increased 19.6% from $44,875,928 2020 to $53,664,571 2023.
Revenue per Employee: Increased 3.47% from Pre-AI: $98,412. Post-AI: $101,830.
ROI:
- **Investment:** $1.5M.
- **Annual Benefits:** $6M.
- **2-year ROI:** ($6M − $1.5M) / $1.5M = 300%.
- **Payback:** 3 months.

Gamification: Making Adoption a Blood-Pumping Sport
The most electrifying AI adoptions transform learning into championship competition, leveraging research showing that gamified approaches can spike engagement by a mind-blowing 60% and improve skill retention by an incredible 40%. When the stakes feel like the World Cup final, people don't just push harder—they push with the intensity of Olympic champions, think faster than supercomputers, and take pride in every victory like it's their first-born child.

High-octane formats include AI hackathons that put cross-functional dream teams on the clock to solve real-world challenges with life-or-death urgency, simulation races where employees test-drive AI in realistic scenarios that feel like video games, and leaderboards tracking departments with the most AI-driven improvements that spark competitive fires hotter than volcanic lava.

Hands-on experience serves as the steering wheel in this gamified rocket ship. McKinsey reports that experiential learning doesn't just boost AI adoption rates—it launches them into orbit by 75% compared to watching paint-dry passive instruction. When employees move from watching AI from the cheap seats to gripping the wheel themselves, anxiety doesn't just stall out—it gets obliterated, competence takes the wheel, and pride crosses the finish line at light speed.

Proof It Works — Real Companies, Real Acceleration, Real Profits

AI-Driven Creative Skills Transformation Case Study: Laing O'Rourke (Large Enterprise)

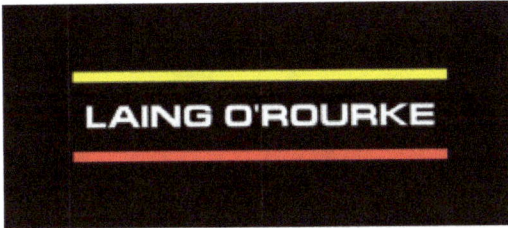

Photo from Laing O'Rourke website

Laing O'Rourke, founded in 1978 by Ray O'Rourke, is a global leader in engineering, construction, and manufacturing, specializing in large-scale infrastructure and building projects. Headquartered in Dartford, Kent, United Kingdom, with significant operations in Sydney, Australia, Laing O'Rourke is known for pioneering technologies like Building Information Modelling (BIM) and prefabrication. Prior to accelerating AI implementation in 2021, the company had approximately 5,607 employees and estimated annual revenue of $4.2 billion USD in 2020.

Problem
Laing O'Rourke struggled with persistent productivity challenges in construction, a sector notorious for inefficiencies. Manual processes in project management, resource allocation, and design optimization led to delays, cost overruns, and suboptimal material usage. The company needed to enhance data-driven decision-making and streamline workflows to remain competitive against rivals adopting advanced technologies, while ensuring employee engagement in a traditionally slow-to-innovate industry.

Challenge to Implementation
Implementing AI across Laing O'Rourke's complex operations faced significant hurdles:

1. **Cultural Resistance**: Construction workers and managers, accustomed to traditional methods, viewed AI as a threat to jobs or an overly complex addition, leading to skepticism.

2. **Skill Gaps**: Many of the 5,607 employees lacked AI literacy, complicating adoption of tools like predictive analytics and generative AI.

3. **Data Integration and Quality**: Fragmented data from project sites and legacy systems posed integration challenges, with concerns about data accuracy and security.

4. **Scalability and Cost**: Scaling AI for diverse projects while justifying costs in a capital-intensive industry was difficult, especially for a middle market firm transitioning to tech-driven operations.

AI People Solution Implementation
Laing O'Rourke accelerated AI adoption by integrating gamification into its strategy, using engaging, game-based elements to foster employee buy-in and streamline training. This approach, combined with communication and storytelling, transformed the workforce's perception of AI. The methodology drew from gamification frameworks and agile development, tailored to construction's unique needs.

Detailed Solution:
1. **Gamification for Engagement**: Laing O'Rourke developed an internal gamified platform, the "Data Academy," in partnership with Multiverse, where employees earned points, badges, and leaderboard rankings for completing AI training modules and applying AI tools on projects. Challenges like "Optimize a Floor Plan" used real project data to gamify tasks like predictive analytics for material usage.

2. **Storytelling and Communication**: Success stories, such as AI reducing project delays by 20% on a London build, were shared in workshops and site briefings to inspire adoption. These narratives framed AI as a "site assistant," enhancing human expertise.

3. **In-Person Events**: Hosted hackathons and "AI Innovation Days" at project sites, where teams competed to solve real-world challenges (e.g., scheduling optimization) using AI tools. These events built camaraderie and hands-on experience, reducing resistance.

4. **Addressing Challenges**:

a. **Resistance**: Gamification made learning fun, with 300 employees enrolled in the Data Academy, earning rewards for milestones, shifting perceptions from threat to opportunity.

b. **Skill Gaps**: Bite-sized, game-based training modules upskilled 90% of participants, with leaderboards fostering friendly competition.

c. **Data Integration**: Partnered with KOPE to digitalize data for AI tools like lattice floor system optimization, ensuring compatibility with BIM.

d. **Scalability and Cost**: Phased pilots on projects like The Whiteley in London validated AI benefits, justifying further investment.

Organizational Impact and Results

Gamification transformed Laing O'Rourke's AI adoption, with 80% of trained employees actively using AI tools within six months. Key outcomes included:

- **Operational Efficiency**: AI-driven design processes accelerated by 25%, and material waste reduced by 15% through optimized layouts.

- **Employee Engagement**: Gamified training boosted participation, with 300 staff enrolled in the Data Academy, and employee satisfaction with tech tools reached 85%.

- **Project Delivery**: Predictive analytics cut project delays by 20%, and augmented reality (AR) via Trimble Connect improved on-site accuracy, enhancing client outcomes.

- **Cultural Shift**: Gamification fostered a data-driven culture, with 70% of employees contributing AI-driven solutions, breaking down silos.

Financial Performance
Revenue Growth: Increased 60% from $2.5B 2020 to $4.0B 2024.
Revenue per Employee: Increased 78.8% from Pre-AI: $210,970. Post-AI: $377,251.

ROI:
- **Investment:** $150M.
- **4-year ROI:** 350%.
- **Payback:** 2 months.

Victory Lap: Owning the Podium Together

The organizations that don't just reach the AI podium but absolutely dominate it aren't just those with the fastest algorithms—they're the ones with communication so clear it could cut diamonds, narratives so inspiring they make movies jealous, and leadership that keeps both technological and human engines tuned for championship-level performance. Successful AI transformation requires a championship dream team approach where metrics and dashboards maintain official victory standings, face-to-face communication serves as trust-refueling pit stops that could power rocket ships, storytelling provides direction and purpose with GPS precision, and gamified experiences sharpen skills and confidence until they're razor-sharp.

When these elements work together like a perfectly tuned race car, organizations don't just finish the race—they obliterate the competition and own the podium like conquering heroes. The victory belongs not just to the CEO or project leads, but to every single employee who contributed blood, sweat, and tears to the transformation. Every customer service champion who mastered the AI chatbot like a ninja, every warehouse warrior who optimized inventory predictions like a fortune teller, and every data analyst who discovered insights that changed everything stands on that podium, basking in collective triumph that makes Super Bowl victories look like participation trophies. In the AI race, finishing isn't enough—the goal is to finish first while ensuring the entire team shares in the glory like championship athletes.

Because ultimately, the podium isn't just a platform for individual achievement — it's thunderous proof that when technology, people, and leadership fire on all cylinders like a perfectly tuned racing engine, the result is something so spectacular that competitors can only watch from below, wondering how they got so thoroughly demolished.

The road to the AI podium is paved with stories that make hearts race, fueled by engagement that burns hotter than rocket fuel, and driven by the collective power of organizations that understand transformation isn't just about technology — it's about unleashing the championship potential that lives inside every human being, waiting to explode into victory.

Proof It Works — Real Companies, Real Acceleration, Real Profits

Hyland Software (Middle Market)

Hyland
creator of OnBase®

Photo from Hyland website

Hyland Software, founded in 1991 by Packy Hyland Jr., is a leading provider of content services and enterprise content management (ECM) solutions. Initially focused on document imaging for healthcare, Hyland evolved into a global leader with its OnBase platform, serving industries like financial services, insurance, government, and higher education. Headquartered in Westlake, Ohio, Hyland is recognized for its customer-centric approach and innovation. Prior to accelerating AI implementation in 2021, Hyland had 4,300 employees and estimated annual revenue of $650 million in 2020.

Problem
Hyland faced challenges in scaling its content services to meet growing demand for real-time data processing and personalized customer experiences. Manual document management and customer support processes were labor-intensive, leading to delays in service delivery and increased operational costs. As competitors adopted AI-driven automation, Hyland risked losing market share without faster innovation and broader accessibility for non-technical users.

Challenge to Implementation
Implementing AI across Hyland's operations presented several hurdles:

1. **Cultural Resistance**: Employees, particularly in customer support and content management, feared AI would replace human roles, leading to skepticism and low adoption.

2. **Skill Gaps**: Many of the 4,300 employees lacked AI expertise, complicating the deployment of tools like intelligent document processing and predictive analytics.

3. **Integration Complexity**: Merging AI with the OnBase platform and legacy systems raised concerns about data security and compatibility.

4. **Stakeholder Alignment**: Coordinating internal teams and external clients on AI's value required clear, compelling communication to overcome skepticism.

AI People Solution Implementation
Hyland accelerated AI adoption by dominating communication and storytelling through internal campaigns, press releases, and client engagements, framing AI as a collaborative enabler. The methodology combined Kotter's 8-Step Change Model for cultural transformation and agile development for iterative AI deployment, emphasizing compelling narratives to drive engagement.

Detailed Solution:
1. **Internal Communication Campaigns**: Hyland launched the "AI as Your Co-Pilot" campaign, using town halls, newsletters, and videos to share success stories, such as AI reducing document processing time by 40%. Leaders like CEO Bill Priemer emphasized AI's role in enhancing human work, reducing resistance.

2. **Storytelling in Press and Events**: Press releases highlighted AI-driven wins, like the Hyland Intelligent Document Processing (IDP) solution streamlining insurance claims, earning media coverage and client trust. Hyland's annual CommunityLIVE conference showcased AI success stories, engaging thousands of clients with live demos.

3. **Employee Engagement through Storytelling**: Internal workshops featured employee testimonials, such as support staff using AI chatbots to resolve queries faster, making AI relatable. Storytelling framed AI as a tool for "unlocking human potential," boosting adoption.

4. **Addressing Challenges**:

    a. **Resistance**: Monthly "AI Success Stories" sessions and leadership endorsements shifted perceptions, with 85% of employees embracing AI tools post-campaign.

    b. **Skill Gaps**: Hyland invested $2 million in AI training for 80% of staff, using storytelling to make complex concepts accessible.

    c. **Integration**: Agile sprints integrated AI with OnBase, with pilot projects ensuring compatibility and security.

    d. **Stakeholder Alignment**: Client webinars and press kits showcased AI's ROI, aligning external stakeholders.

Organizational Impact and Results

Hyland's strategic communication and storytelling transformed AI adoption, achieving:

- **Operational Efficiency**: AI-driven IDP reduced document processing time by 40% and customer query resolution by 50%.

- **Employee Engagement**: 85% of employees adopted AI tools, with satisfaction reaching 90% due to relatable storytelling.

- **Client Trust**: Press and event narratives boosted client adoption of AI features by 60%, enhancing market position.

- **Innovation Culture**: Cross-functional teams developed new AI features, like predictive analytics for compliance, strengthening Hyland's leadership in ECM.

Financial Performance

Revenue Growth: Increased 20% from $650M 2020 to $780M 2023.
Revenue per Employee: Increased 56.4% from Pre-AI: $151,163. Post-AI: $236,363.
ROI:
- **Investment:** $30M.
- **Annual Benefits:** $230M (rev $130M + $100 cost savings).
- **3-year ROI:** ($230M − $30M) / $30M = 667%.
- **Payback:** 3 months.

Victory Lap: Owning the Podium Together

Hyland's AI transformation exemplifies dominating the podium through masterful communication and storytelling. By crafting narratives that resonated internally and externally, Hyland turned skepticism into enthusiasm, aligning 4,300 employees and thousands of clients around a shared AI vision. Every team member — from support staff to developers — stood on the victory podium. This collective triumph, backed by a 667% ROI and 20% revenue growth, proves that when communication, storytelling, and leadership fire on all cylinders, organizations don't just compete — they demolish the competition and redefine industry leadership.

Bottom Line: Winning the AI race isn't about who has the fastest algorithm or the flashiest tech—it's about the synergy between people, process, and performance. The true podium finishers treat communication as the ignition, storytelling as the throttle, and leadership as the finely tuned steering system guiding the team through every high-speed turn. AI adoption is more than a technical rollout — it becomes a shared mission, a championship quest where every employee knows their role and feels their contribution fuels the win. From the shop floor to the C-suite, the organizations that dominate aren't just deploying AI — they're using it to inspire, unite, and accelerate their entire workforce toward collective, market-crushing victory.

Tips for Reaching the AI Podium

☑ **Start with the Green Flag**
Kick off AI with clear, motivating communication.

☑ **Fuel with Storytelling**
Inspire with stories that show AI's benefits.

☑ **Reframe Fear**
Turn worry into confidence by framing AI as a partner.

☑ **Customize the Narrative**
Tailor AI messaging to each team.

☑ **Show Real Wins Early**
Prove AI's value with quick successes.

☑ **Design Victory Dashboards**
Use visual tools to track and celebrate progress.

☑ **Make Pit Stops Count**
Build trust through in-person engagement.

☑ **Gamify Adoption**
Use competition to boost AI learning and use.

☑ **Lead Like a Driver, Not a Mechanic**
Focus on vision while empowering innovation.

☑ **Celebrate the Podium Together**
Share recognition across the whole team.

Chapter 7: Bottom Line

Every corporation is in a race to capture the $4.4 trillion AI jackpot looming by 2030.Buckle up and press the pedal to the metal! AI has hijacked the racetrack, turbocharging profits and leaving laggards choking on dust! Generative AI, the ultimate nitro boost, is rewriting the rules, slashing costs, and igniting 10%–38% profit surges across every industry. From warehouses buzzing with autonomous robots to hospitals crafting factory-precision longevity, AI is no showroom gimmick—it's the high-octane engine driving Industry 4.0.

But the real horsepower? **Culture**.
Companies that rewire their teams to think like CEOs, empower front-liners, and embrace AI as a co-driver are scorching the competition. Buckle up, because this race rewards the bold—those who scale automation, unleash creativity, and steer every employee, from baristas to board members, toward a market-crushing victory.

Here is Your Championship Checklist:

✅ **Run a Culture Audit first**	It is your pre-race inspection to find hidden talent and fix weak spots before hitting the gas.
✅ **Pick your "North Star" track**	AI use cases that align directly with revenue, efficiency, or customer-boosting goals.
✅ **Build your pit crew**	Select which front-liners to equip with AI training to keep the wheels turning mid-race.
✅ **Set safety guardrails**	Protect privacy, reduce bias, and keep human judgment in the driver's seat.
✅ **Start with practice laps**	Launch small pilots to build skill and prove performance before scaling up.
✅ **Watch your telemetry**	Use real-time race dashboards to gauge if teams on-track and in-sync.

✅ **Celebrate every lap**   Share wins, reward contributors, and fuel morale for the next stretch of the race.

These bottom lines aren't just conclusions—they are your victory lap essentials, the proven strategies that separate AI champions.

Bottom Line #1: Culture is the Best Technology Investment

The Reality Check: Start with a Culture Audit—a high-octane, seven-phase X-ray that exposes your strengths, spotlights your blind spots, and rewires mindsets before the next big tech implementation.

Before you spend another dollar on AI tools, invest in the human operating system that will make or break your success. The most sophisticated algorithm can't overcome a culture that resists change, and the simplest AI tool can transform an organization when the culture embraces innovation.

7 Steps of a Culture Audit

✅ **Survey and Collect Leader Data** including staff information

✅ **Interview Leaders and Staff** to determine viewpoints

✅ **Visit Leader Sites** to understand work environments

✅ **Analyze Survey, Interview and Site Data** comprehensively

✅ **Compile and Produce Data Report** with recommendations

✅ **Present Data Report** to executives and middle managers

✅ **Initiate Leadership Development** and targeted coaching

Red Flags That Demand Immediate Culture Audit:

❌ **Staff Disengagement** — Rising turnover and absenteeism

❌ **Customer Stagnation** — Customer satisfaction plateaus

❌ **Growth Stall** — AI spending rises but growth doesn't

114

❌ **Front-line Frictions** — Barriers that slow or hinder front-line staff

❌ **Slow Handoffs** — Delays in passing work along

❌ **Silent Churn** — Customers leave quietly without notice

❌ **Shadow Processes** — Unofficial procedure workarounds

❌ **Costly Rework** — Waste from fixing preventable errors

Bottom Line #2: Every Employee Should "Think Like a CEO"

The Reality Check: The future belongs to companies where everyone's a strategist, a problem solver, and a value creator. The companies that win won't be the ones with the most executives—they'll be the ones with the most CEO-minded employees.

When front-liners think strategically, they don't just execute AI initiatives—they improve them, innovate with them, and evangelize them. This mindset shift transforms AI adoption from a top-down mandate into a bottom-up revolution.

TIPS

✅ **Share the Big Picture** — Over and over again

✅ **Business Literacy** — Better financial decisions occur

✅ **Showcase "Think Like a CEO"** with mentoring/ job shadowing

✅ **Train Every Employee to ask**:
"What would I do, with AI, if I were the CEO?"

✅ **Reward Strategic Thinking and experimentation**

Bottom Line #3: Human Creativity Wins in the Driver's Seat

The Reality Check: Technology alone doesn't win the race. The future belongs to organizations that stop treating creativity as a side dish and start serving it as the main course. AI may provide the horsepower, but human spark decides who takes the checkered flag.

When organizations unlock front-liner superpowers, AI isn't just a tool—it becomes a multiplier. The magic happens when human imagination meets artificial intelligence, creating solutions that neither could achieve alone.

TIPS

✅ **Make AI Training Human-Centric,** shared with co-drivers

✅ **Sanction Failure** — Innovation requires play & safe-to-fail space

✅ **Reward & Incentivize Curiosity & Imagination** — What gets rewarded gets repeated

✅ **Empower Citizen Developers** — Democratize AI creation

✅ **Integrate Creativity with Metrics** — Measure what matters

✅ **Create Playbooks to Scale Culture** with Technology

Bottom Line #4: Communication and Storytelling Are Your Competitive Advantage

The Reality Check: Winning the AI race isn't about who has the fastest algorithm or the flashiest tech—it's about the synergy between people, process, and performance. The true podium finishers treat communication as the ignition, storytelling as the throttle, and leadership as the finely tuned steering system guiding the team through every high-speed turn.

AI adoption is more than a technical rollout—it becomes a shared mission, a championship quest where every employee knows their role and feels their contribution fuels the win. From the shop floor to the C-suite, the organizations that dominate aren't just deploying AI—they're using it to inspire, unite, and accelerate their entire workforce toward collective, market-crushing victory.

Your Victory Framework:
- **Communication** = The ignition system that starts your engines
- **Storytelling** = The throttle that controls your acceleration
- **Leadership** = The steering system that navigates every turn

TIPS for Reaching the AI Podium
These are reminders of the most communication and storytelling tips

for motivating AI championship performance:

- ✅ **Start with the Green Flag**
  Kick off AI with clear, motivating communication

- ✅ **Fuel with Storytelling**
  Inspire with stories that show AI's benefits

- ✅ **Reframe Fear**
  Turn worry into confidence by framing AI as a partner

- ✅ **Customize the Narrative**
  Tailor AI messaging to each team/ department

- ✅ **Show Real Wins Early**
  Prove AI's value with quick successes

- ✅ **Design Victory Dashboards**
  Use visual tools to track and celebrate progress

- ✅ **Make Pit Stops Count**
  Build trust through in-person engagement

- ✅ **Gamify Adoption**
  Use competition to boost AI learning and use

- ✅ **Lead Like a Driver, Not a Mechanic**
  Focus on vision while empowering innovation

- ✅ **Celebrate the Podium Together**
  Share recognition across the whole team

The Champion's Mindset: Your Final Lap Just as no driver begins a race without a detailed course map and clear team strategy, no organization should launch an AI project without a clear communication plan and projected stories that align with company goals.

Your AI transformation isn't just about technology—it's about unleashing human potential at scale. When you get culture right, empower strategic thinking, nurture creativity, and master communication, AI doesn't just improve your operations—it accelerates your entire organization toward market-dominating victory.

You have Re-Wired Culture to Unleash Your Inner AI CEO. Welcome to the podium. The Trophy of wild profits is yours.

CRACK THE CODE OF HUMAN PERFORMANCE

What if you could see inside the minds of your workforce - spotting hidden motivators, early burnout signals, and untapped leadership potential - while exposing whether your company culture is fueling success or driving people out the door?

Behavior Informatics

- X-ray vision into motivation, adaptability, and risk tolerance.
- Predict stars, catch burnout before it spreads
- Surface future leaders.

Culture Audits

- Tear down the glossy slogans.
- Discover the real values, beliefs, and habits shaping daily work and where leadership's message doesn't match reality.

MERGE THE TWO,
and you unlock a GPS for culture + people.

You'll know precisely where employees thrive, where they struggle, and exactly what levers to pull to fix it.

Stop guessing. Start knowing.

This is how tomorrow's companies supercharge performance.

HIRE US!
SEND EMAIL, TODAY!

www.autoworkz.org
ida@autoworkz.org

SenseiiWyze

References

3Cloud Solutions. (2024). Cultivating a culture for AI integration in the workplace. *3Cloud Blog.*

A

Agrawal, A. (2024). Artificial intelligence adoption and system-wide change. *Journal of Economics & Management Strategy,* Wiley Online Library.

AIX - AI Expert Network. (2024). Case study: Bank of America's $4 billion bet on AI. *AIX | AI Expert Network.*

AIX | AI Expert Network. (2024). Case study: Verizon's strategic implementation of AI.

AIX | AI Expert Network. (2023). Case Study: How Nike is Leveraging AI Across its Operations. Retrieved from

Alyve. (2025). Case Studies of Generative AI Adoption in Business.

Automation Workz. (2025a). SenseiiWyze.

Automation Workz. (2025b). Senseii Coaching.

B

Baldoni, J. (2003). *Great Communication Secrets of Great Leaders.* McGraw Hill.

Bank of America. (2024). AI patents at BofA increase 94% since 2022. *Bank of America Newsroom.*

Bank of America. (2025a,). AI adoption by BofA's global workforce improves productivity, client service. *Bank of America Newsroom.*

Bank of America Corporation. (2024). Investor relations. *Bank of America Corporation (BAC).*

Bar-On, R. (2006). The Bar-On model of emotional-social intelligence (ESI). *Psicothema,* 18, 13-25.

Booz Allen Hamilton. (2025). Change management for artificial intelligence adoption. *Booz Allen Insights.*

Boyatzis, R. (2019). *Helping people change: Coaching with compassion for lifelong learning and growth.* Harvard Business Review Press.

Brackett, M. A., Rivers, S. E., & Salovey, P. (2011). Emotional intelligence: Implications for personal, social, academic, and workplace success. *Social and Personality Psychology Compass,* 5(1), 88-103.

Brookings. (2024). Generative AI, the American worker, and the future of work.

Brynjolfsson, E., & McAfee, A. (2014). *The Second Machine Age: Work, Progress, and Prosperity in a Time of Brilliant Technologies.* W.W. Norton & Company.

Building Point UK & Ireland. (2022). Augmented Reality Case Study - Laing O'Rourke & Trimble.

Bullfincher. (2024). Cigna Corporation number of employees 2018-2024.

Bureau of Economic Analysis (BEA). (2022). Corporate Profits After Tax (without IVA and CCAdj), Annual. U.S. Department of Commerce.

Business Insider. (2025). A Wholesale Gown Distributor Is Using AI for Its E-Commerce Operations. Retrieved from

C

Caterpillar Inc. (2024). Caterpillar Chairman and CEO Jim Umpleby to become Executive Chairman; Caterpillar Chief Operating Officer Joe Creed Elected as Caterpillar's Next CEO. *Corporate Press Release.*

Caterpillar Inc. (2024). Caterpillar Reports Fourth-Quarter And Full-Year 2023 Results. *Corporate Press Release.*

Caterpillar Inc. (2024). Future of AI at Caterpillar. *Corporate News.*

Caterpillar Inc. (2025). Company History Timeline.

Caterpillar Inc. (2025). D. James (Jim) Umpleby III. *Corporate Leadership Profile.*

Caterpillar Inc. (2025). Meet the Founders. *Company History.*

Caterpillar Inc. (2025). The Best Holt and Best Story You've Never Heard. *Corporate History.*

Cause IQ. (2025). GreenPath Incorporated | Farmington Hills, MI

Change Management Institute. (2024). Engaging change communication with storytelling. *Change Management Institute.*

Change Management Insight. (2023). Change management storytelling examples, types, and techniques. *Change Management Insight.*

Chen, L., & Rodriguez, M. (2023). Leadership analytics in e-commerce environments: Predictive insights for management effectiveness. *Journal of Business Analytics,* 15(3), 234-251.

Chen, L., & Wang, S. (2020). Machine learning applications in online education: The VIPKid case study. *Journal of Educational Technology Research,* 45(3), 234-251.

Chen, L., Rodriguez, M., & Thompson, K. (2023). Emotion AI at work: Implications for workplace surveillance, emotional labor, and emotional privacy. *Proceedings of the ACM Conference on Computer-Human Interaction,* 15(3), 145-162.

Cherniss, C. (2010). Emotional intelligence: Toward clarification of a concept. *Industrial and Organizational Psychology,* 3(2), 110-126.

Cigna Health Benefits. (2024). Innovating health care.

Cigna Newsroom. (2025). Top health care trends of 2025 and how they will impact U.S. employers.

Citizen Musings. (2025). The power of storytelling in organizational change: Inspiring behavioral shifts through relatable narratives. *Medium.*

CNBC. (2024, March 6). Just 38% of employers provide AI training to workers. Retrieved from

Colvin, G. (2015). *Humans Are Underrated: What High Achievers Know That Brilliant Machines Never Will.* Portfolio.

Comprend. (2024). AI in corporate communications: current uses and future roles. *Comprend Insights.*

CTO Magazine. (2025, January 8). Inside Bank of America AI: 90% adoption across workforce. *CTO Magazine.*

CultureManagers. (2025). 6 Metrics to Measure the Success of Culture-Building Efforts.

CXOTalk. (2025, May 18). Technology and Innovation Strategy with the CTO of Caterpillar.

Crunchbase. (2021). VIPKID company profile: Funding, valuation & investors.

Crunchbase. (2025). Automation Workz - Crunchbase Company Profile & Funding.

D

D3.harvard.edu. (2023). Sprinting into the Future: Nike's AI Strategies for Tomorrow.

D CEO Magazine. (2024, December 5). How Jim Umpleby Led Global Construction Giant Caterpillar to Record Financial Highs.

D CEO Magazine. (2024, December 17). Why D CEO Named Jim Umpleby Its 2024 CEO of the Year.

Dauntless Agency. (2024, March 18). Driving Growth—Caterpillar's Blueprint for Digital Transformation Success.

Davenport, T. H., & Bean, R. (2024). Mayo Clinic's healthy model for AI success. *MIT Sloan Management Review.* Retrieved from

Davenport, T. H., & Kirby, J. (2016). *Only Humans Need Apply: Winners and Losers in the Age of Smart Machines.* Harper Business.

Deloitte Insights. (2022). The gamification of learning and development. *Deloitte University Press.*

Deloitte US. (n.d.). AI transformation and culture shifts. *Deloitte Insights.*

designveloper.com. (2025). Nike Digital Transformation: A Case Study of Modern Commerce. Retrieved from

Digital Defynd. (2025). 5 ways Ford is using AI [Case Study] [2025].

Digital Defynd. (2025). 7 Ways Nike is Using AI [Case Study] [2025].

E

Edmondson, A. C. (2019). *The fearless organization: Creating psychological safety in the workplace for learning, innovation, and growth.* Wiley.

eGain. (2024). GreenPath and eGain Expand Capabilities Offered by Anytime-Anywhere Virtual Financial Coach

eGain. (2021). Pandemic Imperative for Insurance: Virtual Financial Coaching.

El Kaliouby, R. (2020). *Girl decoded: A scientist's quest to reclaim our humanity by bringing emotional intelligence to technology.* Currency Books.

Emberin. (2024, May). Conducting a diversity audit: Ensuring inclusivity in the workplace. *Emberin.*

EqualOcean. (2020, August 18). VIPKID records positive unit operating profit in 1H 2020. *EqualOcean Insights.*

EW Group. (2024, February). What a diversity audit is, the benefits, how to start one. *EW Group.*

EY. (2024). How Caterpillar is using technology on its journey to improve financial forecasting.

F

Firstup. (2024). Leveraging generative AI in corporate communications: Overcoming challenges and maximizing benefits. *Firstup Blog.*

Forbes. (2023). Why Transparent Communication Is Vital for AI Implementation.

Forbes. (2025). Carhartt | Company Overview & News.

Forbes. (2025). Nike | Company Overview & News.

FullStack Labs. (n.d.). Assessing AI readiness: A practical guide for companies. *FullStack Labs.*

G

Gallup. (2023). *State of the global workplace 2023 report.* Gallup Press.

Gallup. (2025). Successful organizational change needs a strong narrative. *Gallup Workplace.*

Gannon, P. (2001). *Inside Room 40: The codebreakers of World War I.* Ian Allan Publishing.

Gartner. (2021). Gamification Strategies for Workforce Engagement.

Genesys. (2019). Mapping your organization's internal stakeholders for AI ethics success. *Genesys Blog.*

Globe Newswire. (2024). The Cigna Group Digital Transformation Strategies Report 2024 - Accelerators, Incubators and Other Innovation Programs

Goleman, D. (1995). *Emotional intelligence: Why it matters more than IQ.* Bantam Books.

Goleman, D. (2020). *Emotional intelligence: Why it matters more than IQ (25th anniv. ed.).* Bantam Books.

Goleman, D., & Boyatzis, R. (2022). Emotional intelligence and AI adoption in organizations. *Harvard Business Review,* 100(3), 78–87.

Gonzalez, A., & Martinez, S. (2024). AI behavioral analysis: The application of artificial intelligence to analyze, predict, and influence human behavior. *HSE AI Research Journal,* 8(2), 78-95.

GP Strategies. (2024). AI's impact on storytelling: Can it replicate human experiences? *GP Strategies Blog.*

Grant Thornton & Oxford Economics. (2019). Return on Culture: Correlating Culture with Business Outcomes (survey-based study of U.S. organizations).

Great Place To Work. (2025). How High-Trust Cultures Boost Productivity and Increase Revenue per Employee.

GreenPath Financial Wellness. (2021). 2021 Annual Report.

Growjo. (2025). Hyland Software: Revenue, Competitors, Alternatives.

H

Harvard Business Review. (2022). Emotional Intelligence and Retention.
Harvard Business Review. (2023). The power of gamification in employee engagement. *Harvard Business Review Press.*
Harvard Business Review. (2023). Storytelling that drives bold change. *Harvard Business Review.*
Harvard Business Review. (2024). Three Steps to a High-Performance Culture (ANZ case study).
Harvard Business School. (2018, November 13). VIPKID: Machine learning in online education [Case Study]. *Harvard Business School Case Collection*, Case 9-819-025.
Healthcare Dive. (2025, June 12). Cigna launches new generative AI assistant for members. Retrieved from
HR Executive. (2023). What's Keeping HR Up at Night? [Report].
HR Executive. (2023). *Data visualization and employee engagement: 2023 trends report.* HR Executive Media.
Hyland. (2023). Hyland Accelerates AI Innovation with OnBase

I

IBM Institute. (2025). Transforming change management with responsible AI. *IBM Think Insights.*
iexpert.network. (2023). Case Study: How Nike is Leveraging AI Across its Operations.
Inside Small Business. (2024). "Case studies: how SMEs are using AI to compete with big players."
Insight7. (2025). AI tools that turn emotion analytics into strategic dashboards. *Insight7.*
International Institute for Learning. (2025). The impact of artificial intelligence on stakeholder relations management practices. *IIL Blog.*
ISM Guide. (2025). Caterpillar Pioneers the Use of XR, VR, AR, and AI in Its Business Today.

K

Kneller, T. (2023). The magic of storytelling in change management. *Medium*
KOPE. (2024). Responding to project bids at speed - Case Study.
Kotter, J. P. (2012). *Leading Change.* Harvard Business Review Press.
Kotter, J. P. (2014). *Accelerate: Building Strategic Agility for a Faster-Moving World.* Harvard Business Review Press.
Kumar, A., & Singh, P. (2023). E-commerce AI integration methodologies: A systematic review of implementation frameworks. *International Journal of Retail & Distribution Management*, 51(8), 1123-1140.

L

Laing O'Rourke. (2024). Annual Report and Accounts 2023.
Lantern Studios. (2024). The importance of organizational and cultural readiness for AI success. *Lantern Studios.*
Leeman, R. (2018, August 2). Change management and storytelling. *LinkedIn.*
LinkedIn. (2025). How Storytelling Transformed Nike's Brand and Drove Massive Results.
Little, J. (2014). *Lean Change Management: Innovative Practices for Managing Organizational Change.* Happy Melly Express.
LumApps. (2025). Understanding AI in corporate communication: best practices. *LumApps Insights.*

M

MacroTrends. (2024). Bank of America revenue 2010-2025 | BAC. *MacroTrends.*
MacroTrends. (2025). Ford Motor Company - total number of employees 2015-2024.
Mayer, J. D., & Salovey, P. (1997). What is emotional intelligence? In P. Salovey & D. Sluyter (Eds.), *Emotional development and emotional intelligence: Educational implications* (pp. 3-31). Basic Books.
Mayo Clinic. (2024). Artificial intelligence, data science and informatics
McKinsey & Company. (2021). The COVID-19 recovery will be digital: A plan for the first 90 days.
McKinsey & Company. (2023). *The Economic Potential of Generative AI: The Next Productivity Frontier.*
McKinsey & Company. (2023). *The state of AI in 2023: Generative AI's breakout year.* McKinsey Global Institute.
McKinsey & Company. (2024). The impact of AI on logistics and supply chain management. Retrieved from McKinsey Global Institute.
McKinsey & Company. (2025). AI in the workplace: A report for 2025. *McKinsey Insights.*
McKinsey & Company. (2025). How AI is transforming strategy development. *McKinsey Digital.*
McKinsey & Company. (2025). The state of AI: How organizations are rewiring to capture value. *McKinsey Global Institute.*
MDPI. (2024). Methodological approach to assessing the current state of organizations for AI-based digital transformation. *MDPI Open Access Journals.*
Medium. (2024). The benefits and challenges of AI in corporate communication. *Medium Articles.*
Mehrabian, A. (1971). *Silent messages: Implicit communication of emotions and attitudes.* Wadsworth Publishing.
Mi, C. (2020, June 24). Building the future of education through AI innovation. *VIPKid Blog.*
Microsoft Cloud Blog. (2024, November). Assess your AI readiness. *Microsoft Azure Blog.*
MIT Sloan Management Review. (2025). Building an organizational approach to responsible AI. *MIT Sloan.*
MIT Sloan Management Review. (2025). The enduring power of data storytelling in the generative AI era. *MIT Sloan.*
Multiverse. (2021). How Laing O'Rourke is building its workforce for a data-driven future.

N

National Center for Biotechnology Information. (2024). The first generative AI prompt-a-thon in healthcare: A novel approach to workforce engagement with a private instance of ChatGPT. *PMC*.

National Center for Biotechnology Information. (2024). Prompt engineering as an important emerging skill for medical professionals: Tutorial. *PMC*. Retrieved from

New Mexico State University. (2025). 7 smart ways AI is transforming PR & communication strategies. *NMSU Global Blog*.

Nike, Inc. (2018). FY18 Annual Report.

Nike, Inc. (2023). FY23 Annual Report.

P

Park University. (2025). Leading through change: Communication strategies for organizational transformation. *Park University Blog*.

Pitchgrade.com. (2024). Nike, Inc.: AI Use Cases 2024.

PitchBook. (2021). VIPKid company profile: Valuation, funding & investors. *PitchBook Data Inc.*

PitchBook. (2025). Automation Workz 2025 Company Profile. Retrieved from

PMC. (2024). Diversity impact on organizational performance. *PubMed Central*.

PR Newswire. (2022). Hyland Named a Leader in IDC MarketScape for Content Services Platforms

PRNewswire. (2025, January 30). Caterpillar Reports Fourth-Quarter and Full-Year 2024 Results.

ProPublica. (2025). Greenpath Inc - Nonprofit Explorer. Retrieved from

Prosci. (2023). ADKAR: A Model for Change in Business, Government, and Our Community. *Prosci Learning Center Publications*.

Prosci. (2025). AI in change management: Early findings. *Prosci Research*.

PwC. (2023). *AI and workforce evolution: Building the workforce of the future*. PricewaterhouseCoopers.

PwC. (2023). *The Future of Work Report*.

PwC. (2023). *Sizing the Prize: The Value of AI for Your Business*.

R

Ransbotham, S., Kiron, D., Gerbert, P., & Reeves, M. (2022). The cultural benefits of AI adoption. *MIT Sloan Management Review*, 64(1), 1–10.

Remote. (2025, February). Diversity audits: What they are, benefits, and how they work. *Remote*

RocketReach. (2023). Carhartt Information.

Rosen, M. (2019). AI implementation strategies in EdTech: Lessons from Chinese unicorns. *MIT Technology Review*, 122(4), 78-85.

RSM US. (2025). Middle Market Firms Rapidly Embracing Generative AI, But Expertise Gaps Pose Risks: RSM 2025 AI Survey.

Russell, S., & Norvig, P. (2020). *Artificial intelligence: A modern approach (4th ed.)*. Pearson.

S

Sadaric, A. (2023, January 25). What does aesthetic storytelling have to do with successful organizational change? A leadership capability perspective. *Medium*.

Salovey, P., & Mayer, J. D. (1990). Emotional intelligence. *Imagination, Cognition and Personality*, 9(3), 185-211.

Schein, E. H., & Schein, P. (2017). *Organizational Culture and Leadership (5th ed.)*. Jossey-Bass.

ScienceDirect. (2022, September). Technology readiness and the organizational journey towards AI adoption. *ScienceDirect*.

SHRM. (2023). *Workplace Culture and Innovation Report*.

Society for Human Resource Management. (2025). How organizational culture shapes AI adoption and success: Q&A with Jessica Kriegel of Culture Partners. *SHRM AI Hub*.

Society for Human Resource Management, & Qualtrics XM Institute. (2023). Culture, engagement, and the bottom line [White paper]. *SHRM*.

SPD Technology. (2025). AI in logistics: Use cases, benefits, and challenges. *SPD Technology Blog*.

Sprinklr. (2025). The transformative impact of AI in communication. *Sprinklr Blog*.

SS&C Blue Prism. (2025). AI change management – Tips to manage every level of change. *Blue Prism Resources*.

Stanford Social Innovation Review. (2025). Changing popular narratives around AI to empower workers. *SSIR*.

T

Tectonic. (2025). Cigna enhances member experience with AI-powered digital tools.

TechNexus. (2025, April 5). Verizon leverages generative AI to enhance customer experience. Retrieved from

The Cigna Group. (2025, January 30). The Cigna Group reports fourth quarter and full year 2024 results, establishes 2025 outlook and increases dividend.

Thompson, D., Martinez, S., & Lee, K. (2024). Machine learning applications in healthcare supply chain optimization: Ensemble methods and predictive analytics. *Supply Chain Management Review*, 28(4), 445-461.

TM Forum. (2022). Verizon uses AI and machine learning to optimize supply chain inventory.

Tracxn. (2025). Carhartt - 2025 Company Profile, Team & Competitors.

V

van der Steen, M., Johnson, P., & Williams, R. (2022). Story-making to nurture change: Creating a journey to make transformation happen. *Journal of Knowledge Management*, 26(4), 892-915

Verizon. (2025). Upskilling and reskilling programs for digital skill training. Retrieved from

Verizon. (2025). Verizon unveils AI strategy to power next-gen AI demands. Retrieved from

Verizon Business. (2024). 3 ways technology will help enterprises navigate a challenging 2024. Retrieved from

Ververica. (2021). Case study: How VIPKid processes real-time data at scale.

VIPKid. (2021, March 23). How AI empowers teachers in online education. *VIPKid Educational Research Journal*, 8(2), 15-28.

Vorecol. (n.d.). The role of AI in enhancing organizational culture assessment tools. *Vorecol*.

Voltage Control. (2024). Adopting AI-driven change management: Key strategies for organizational growth. *Voltage Control Articles*.

Y

Yahoo Finance. (2025, February 10). Ford Motor full year 2024 earnings: EPS beats expectations, revenues lag. *Yahoo Finance*.

Yukl, G. (2013). *Leadership in organizations (8th ed.).* Pearson.

Z

Zhang, H., Liu, Y., & Brown, J. (2021). Employee empowerment in AI adoption: A comparative analysis of Chinese technology companies. *Strategic Management Journal*, 42(7), 1456-1478.

ZoomInfo. (2025). Automation Workz - Overview, News & Similar companies.

ZoomInfo. (2025). Hyland Software - Overview, News & Similar Companies.

www.ingramcontent.com/pod-product-compliance
Lightning Source LLC
Chambersburg PA
CBHW071428210326
41597CB00020B/3707